AF531306

EUCALYPTUS:
ENDURING MYTHS, STUNNING REALITIES

By

S.A. Abbasi

PhD, DSc, FNASc, FIIChE, FIE
Senior Professor & Director
Centre for Pollution Control and Energy Technology
Pondicherry (Central) University, Kalapet
Pondicherry 605 014

N. Ramesh, *MS*

Department of Science, Technology, and Environment
Government of Pondicherry

S. Vinithan, *M.Sc, PhD*

Senior Environmental Consultant
Pondicherry

DISCOVERY PUBLISHING HOUSE
NEW DELHI

First Published–2004

ISBN: 81-7141-892-9

Published by:

DISCOVERY PUBLISHING HOUSE

4831/24, Prahlad Street, Ansari Road, Darya Ganj
New Delhi–110 002 (India)
Phone: 23279245, • Fax: 91-11-23253475
e-mail: dphtemp@indiatimes.com

Printed at:

ARORA OFFSET PRESS
Laxmi Nagar, Delhi-92.

Dedicated to

The Composer:

Daadoo

(Prof. Mahesh Chandra Chattopadhyaya, University of Allahabad)

The Singer:

Prashant

(Prof. P.K. Bhattacharya, IIT Kanpur)

...and the sweet music they created

—Shahid

Preface

Eucalyptus has touched the lives of most of us in one way or the other. To some it has done so, literally, in the form of eucalyptus oil (also called Nilgris oil). To some others it has done so in the form of paper made from the tree's pulp. To some others it has meant building material, and to some a wayside view, if nothing else.

The lanky, upright, tree is often seen standing on soil where other trees refuse to grow. The tree has been a favourite of farmers looking for quick returns, of social forestry officers for the certainty of its success, and, of course industrialists.

But eucalyptus has its detractors and these detractors have been so obstreperous that a storm has been raised by them, which continues to rage. According to it's baiters, eucalyptus is a harmful, cancerous presence; and those who support eucalyptus are enemies of ecology and the nation.

In the midst of all the hype we have made a modest attempt to find the truth. This book is a result of the attempt spanning five years. Even as the focus of the book is eucalyptus, many things said in the book may appear to transcend the immediate context. The studies described in the book have revealed that it is not eucalyptus which causes ecological damage but the way we overexploit its virtues that does. In this central point may well be lying the entire philosophy of environmental management. Nothing is bad, nothing is harmful; it is we who invite harm by misusing right things in the wrong manner. When stated in these simple terms it looks rather straightforward, but it is one truth the mankind keeps overlooking again and again.

We have tried our best to proceed with an open mind and each aspect of the eucalyptus controversy has been examined by us dispassionately and solely on the basis of hard evidence.

We are grateful to the management of *The Indian Forester* for bestowing the coveted S.K. Seth Prize on this work. We thank Ms G. Uma, Mr. S. Karthikeyan, and Ms. Debarati Chaudhary for their painstaking assistance. We are beholden—SAA and SV to Pondicherry University and NR to Government of Pondicherry—for consistent encouragement.

Abbasi
Ramesh
Vinithan

Contents

1

Introduction

In a world so desperately short of wood, can a tree that grows fast, thrives in the most adverse of situations, provides biomass that has myriad uses ranging from fuelwood to medicinal to industrial—be anything but an unmixed blessing? Can a tree of which each and every portion is easily utilisable, which provides benefits for the rich and the poor alike, be anything but a boon?

Eucalyptus is one such genera of trees. But at present it is hardly regarded as a gift of God by majority of environmentalists. To them, eucalypts are a curse; a scourge, a harbinger of despair and destruction.

While the supporters of eucalypts find nothing wrong with the tree, the opponents find nothing right! There is no other tree in the world that evokes such strong and conflicting emotions in the proponents of the rival schools of thought. As a result, environmental impacts of eucalyptus plantations, big or small, have become topics of acrimonious discussions on all platforms—scientific or social. Critics of eucalypts point out that they lower the water table, dry up perennial streams, deplete soil moisture, exhaust nutrients, reduce soil fertility, discourage undergrowth, and are susceptable to pest attacks... allegations galore. The supporters of eucalypts refute these charges with the claims that the tree does not cause any of these adverse impacts but is a bonanza for social foresters. These claims and counter-claims are reviewed in detail in chapter 2.

Which of the two sides is on the right? Where does the truth lie?

Eucalypts, natives of Australia, have been taken to countries far and wide by persons anxious to capitalise on the trees' seemingly endless virtues. A typical tree of eucalyptus attains the same height in five years that a casuarina tree does in twenty years (Joyce, 1988). It has been claimed that a eucalyptus tree consumes less than 0.5% water for production of one gram of biomass, an efficiency higher than most of the other genera of tree (Hasankutty, 1986). A typical eucalyptus tree reaches maturity within a decade, and easily coppices. Teak grown for 60 to 80 years in India has a mean annual increment of about 4-8 m^3/ha year whereas yields of *Eucalyptus camaldulensis* plantations in the drier tropics are often about 5-10 m^3/ha year and in wet regions around 30 m^3/ha year over a rotation period of 10-20 years (Evans, 1982)! Whereas eucalypts can be grown at the density of 2,500 to 3,000 trees per hectare, other trees common in India such as *Syzegium cumini, Tamarindus indicus, Azardirachta indica* and *Mangifera indica* even when planted as pure crops can not thrive in populations above 100 per hectare (Shyamsunder, 1986). Eucalypts are comparable to mahogany in terms of the hardness of their wood (Wilson, 1980).

Most eucalyptus species are resistant to drought, frost, heat and alkali (Wilson, 1980). They have been successfully grown over the tropics, sub-tropics and cooler regions (Calle and Hall, 1988). In India *E. tereticornis* has been planted in varying rainfall zones (500-3,500) mm/year), soil types (coastal sand to stiff clay, acidic in pH, 4.5, to alkaline in pH, 9); temperature (tropics with a maximum of 47°C to Himalayan zones with a minimum of 1°C) and topography (river banks to bouldery hillocks; Shanmugam, 1986).

Eucalypts are amazingly multiutility trees. Every part of the trees of this genera has commercial significance. They are among the best—and most utilised—sources of hardwood pulp for paper, rayon and other cellulose products. The bark is used for tanning. The wood is used for making ships, railroad ties, telegraph poles, fencing and piers. The leaves yield aromatic oils which are used as antiseptics, deodorants, astringents, and decongestants. So remunerative are the eucalyptus leaves that the sale value of eucalyptus oil is nearly three times the cost of extracting the oil (Shyamsunder, 1986). Eucalyptus leaves also form the sole food of the Australian Koala, an animal which evolved with them (Joyce, 1988). Eucalypts are planted as windbreaks in strips to protect other crops. They are also planted extensively along roads, canals, railroads; and wastelands such as mine overburden, deforested lands, and other infertile lands in need of reclamation.

The fast growth rate of eucalypts and their multiutility value led pulp producers to introduce them as star performers of social forestry programmes in India and elsewhere (Joyce, 1988). The crisis in fuelwood in the Third World Countries drew the attention of aid agencies, especially the World Bank, and eucalypts became almost automatic choices for them as the quickest solution (Joyce, 1988).

The increasing importance of eucalypts for plantations was indicated in the first edition of *"Eucalyptus for planting"* by FAO in 1954. According to Lamb (1973), the FAO Seminar on tropical pines in Mexico in 1960 more than anything else awakened tropical countries to the value of eucalypts.

In Brazil, largely owing to the pioneering work of de Andrade, several hundred thousand hectares of eucalyptus trees were planted in the 1920s and 1930s notably *E. saligna, E. camaldulensis, E. citriodora* and *E. tereticornis* (Penfold and Willis, 1961).

In India, a eucalyptus species was first introduced as a garden tree way back in 1790 at Nandi Hills in Karnataka State. In 1843, the species, probably *E. globulus,* was planted in the Nilgiri hills, India (Penfold and Willis, 1961; Chaturvedi, 1976).

Eucalypts were brought to the warm coastal regions of Southern India in 1832 (at Madras) where they got established as easily as they had earlier done in the much cooler climate of Nilgiris (Yagnaswami, 1961). Later, from 1956, they were raised in the Nilgiris to meet firewood requirements. Eucalyptus has by now become a widely planted genera in social forestry. Its ability to establish itself successfully on wastelands has earned it a place in soil reclamation programmes, as well. The species commonly seen in India include *E. citriodora, E. tereticornis, E. globulus,* and *E. camaldulencis* (Kedarnath, 1986).

The Present Work

But to the detractors of eucalyptus all the attributes mentioned above are nothing but a 'fatal lure'; they believe that beneath the vaneer of utility lies the ugly face of a 'colonising', 'dictatorial', and 'totalitarian' genera which grow but at the cost of soil, water, climate, and other vegetation. To them eucalypts are harbingers of environmental degradation and devastation. The arguments put forward by them are discussed in some detail in chapter 2.

Where does the truth lie? A very extensive survey of all the past work by us revealed that the controversies concerning eucalypts have arisen and have been sustained because there is no conclusive hard data with which the arguments can be settled one way or the other. All the studies—in support or against eucalypts—have been conducted thus far in plantations which have either faced extensive human interference, or which have been mixed up with other types of natural resource exploitation (such as excessive ground water extraction for uses other than irrigating eucalypts), making it impossible to separate the impacts of eucalypts from other camouflaging factors.

We were fortunate in locating large number of plantations of *Eucalyptus globulus-tereticornis (Eucalyptus hybrid)* and other tree species (Plate 1) in the sprawling campus of Neyveli Lignite Corporation (NLC), 200 km West of Madras. NLC has raised these plantations to counter the adverse effects of deforestation—devegetation caused by their lignite mining activities. The plantations stand on identical types of soil and climatic conditions, and are left undisturbed. It thus became possible to study the environmental impacts of *Eucalyptus globulus-tereticornis (Eucalyptus hybrid)*

Plate1. A typical plantation of *Eucalyptus hybrid* (above) and *Acacia auriculiformis* (below) in the study area.

plantations and compare them with the impacts of other plantations under identical conditions of soil, moisture, and meteorology. The studies, conducted during 1988-1993, encompassed undergrowth diversity, soil moisture, soil nutrients (NPK), soil organic carbon and wi'd life.

Objectives of the Study

The objectives were to study the environmental impacts of *Eucalyptus hybrid* monocultures, monocultures of other tree species, and plantations of mixed cultures to see whether *Eucalyptus hybrid* exerts adverse impacts compared to other mono- and- mixed-culture plantations. The studies involved the following:-

(a) enumeration of undergrowth species found in different types of plantations;

(b) quantification of the undergrowth survey by working out index values for diversity, evenness, richness, and dominance of the undergrowth in the plantations to assess the cohabitability of the afforested tree species with the undergrowth;

(c) preperation of an inventory of the undergrowth species and enlist their habits and utility or nuissance values;

(d) estimation of the soil moisture content, soil organic carbon content and soil nutrient (NPK) content to assess whether *Eucalyptus hybrid* had any impacts different from other tree species.

(e) preparation of checklists of birds and butterfly species;

(f) comparison of the findings of 1988 and 1993 studies, to accurately guage the influence of time on the ecology of the plantations as well as to assess the seasonal variation (temporal influence) on the diversity or richness of the undergrowth.

2

The Eucalyptus Controversy

Introduction

Eucalypts have become the focal point of a raging controversy over the past two decades *vis-a-vis* their impacts on the environment. We have mentioned in chapter 1 that no genera of trees has attracted such intense and acrimonious debate on its perceived ills and virtues as eucalypts. The views for or against eucalypts have become so polarized that at times the criticizm or appreciation is based solely on prejudice than on a balanced consideration of the facts.

Eucalyptus first came to India in 1790 during the reign of Tipu Sultan and established itself as a garden tree in Nandi Hills, Karnataka. This activity was primarily intended to meet the firewood demand. Planting on a large scale in India, however began only around 1856. Along with plantations of teak and the exotic acacias, eucalypts were introduced in Nilgiri hills in 1858 where *E. globulus* was planted along with *E. robusta*. Another species of eucalyptus- *E. citriodora* was introduced in mid 31930 in Karnataka and Kerala. The 1960s heralded the introduction of the hybrid *E. tereticornis-E. globulus* popularly known as *E. hybrid* on a massive scale throughout India from the farthest North to the Southernmost tip.

While more than 2.5 lakh hectares of forest and rainfed farmland have been diverted for eucalyptus plantations in the state of Karnataka, in the North-western States of Punjab and Haryana, eucalyptus plantations account for nearly 1.5 per cent of the total land under cultivation. About 7.5 million hectares of land is currently under eucalyptus plantations in India, accounting for about 8% of the Global coverage.

Why did eucalyptus achieve such a rapid spread? Why has this genera of trees become a favourite of the foresters all over the world? How beneficial is it? Does it have any adverse effects? These are among the questions most frequently asked by ecologists, sociologists, economists, and social foresters.

The spurt in population growth and industrialization during the post-independent years enhanced the demand for wood products in India accentuating the pressure on the forest resources. The demand was far in excess of supply. Several strategies were adopted to increase the output of wood, to meet the industrial and domestic demands. One of the steps was to introduce fast-growing species in plantation forestry which till the mid-50s had been dominated by the slow growing hardwood. This is how eucalyptus, among the fastest growing trees, came into the picture and has since then remained as a favourite in social forestry programmes.

Whatever be the climatic conditions, one or the other species of eucalyptus is always successful in establishing itself. Therefore as a group, the eucalyptus have proven themselves to be extremely successful in a large number of countries. They have been used, for example, as the basis of the massive plantings to supply charcoal to the Brazilian steel industry. In the cooler highlands and inter-Andean valleys of Bolivia, Peru, Equador and ports of Colombia, eucalyptus are the species most sought for fuelwood and other uses. They have been widely planted in tropical Ethiopia and other parts of Africa and, as stated earlier, they have proved to be the most popular species in the farm forestry programmes in India.

There are instances when eucalyptus has been even planted as an effective alternative to sugarcane. One such instance occurred in Nasik, India; when resort to eucalyptus was taken after three successive years of drought had caused a collapse of sugarcane production.

A fresh spurt in eucalyptus plantation has occurred in recent years to overcome the serious scarcity of fuelwood. Forestry experts stress that harvesting of eucalyptus for fuelwood can be done once every four years while there needs to be an interval of eight years before eucalyptus can be harvested for pulpwood.

Overkill

So successful were the initial experiences with eucalyptus that an element of overkill came in. Planters eager to cash in on the virtues of the tree began planting any and every species of the tree even in agroclimatic zones which were totally unsuitable to those species. This has resulted in some well publicised failures. The total destruction of a long stretch of eucalyptus plantation in Kerala, India by the Pink and Blight disease is one such example. Another unfortunate fallout of the high productivity of eucalyptus has been that natural forests have been cleared all over the world to plant it. This had led to an outcry from the environmentalists and social activists.

At present while on one hand official foresters and big farmers have found in the species a panacea for solving the problems of ecorestoration, on the other hand small farmers along with social activist groups all over the country have questioned the preference given to eucalyptus in social forestry programmes. Environmentalists in India have also raised alarm against the expansion of eucalyptus monoculture in the arid and semi-arid regions. The protests gained such a momentum that in March 1983, farmers in Karnataka, India, uprooted eucalyptus seedlings from forest nurseries. In June 1988, farmers in Thailand protested against a eucalyptus tree planting project. In 1989, the chipko activists were arrested for doing what their counterparts did in Karnataka, India. Kerala, India, was also not left far behind.

The controversy has continued to rage because most of the eucalyptus plantations suffer from human interferences and it becomes very difficult to separate the adverse effects of eucalyptus on the environment from other possible causes. This chapter aims to discuss the various elements of the controversy. We have presented the gist of the views of the protagonists as well as the antagonists and then given our own comments on each aspect of the controversy. We have also briefly mentioned how our studies, detailed in subsequent chapters, have covered a major knowledge-gap *viz* extensive long-term experiments on the ecology of undisturbed eucalyptus plantations *in comparison* with the ecology of other equally undisturbed plantations consisting of various 'known to be benign' tree species—in monocultures as well as mixed cultures. Further, these studies have been conducted on plantations grown under identical agro-climatic conditions. The results have thus no danger of being masked or camoflauged by extrenuous factors such as human or cattle disturbance.

Elements of Controversy

Eucalyptus as a Tree Used in Social Forestry

The most extensive use to which eucalyptus has been put is in social forestry. The saplings of eucalypts have a higher survival rate than other trees identified for social forestry- acacia, leucaena, casuarina, prosopis *etc*. The tree grows fast, needs minimal supervision, and provides greater yield of commercially useful biomass per unit land area and per unit time than most other trees.

Forest Departments in India have taken to eucalyptus in a big way because of the attributes mentioned above. For similar reasons the genus eucalyptus has been a tree of choice for non-governmental entrepreneurs, and farmers.

But the strongest opposition to eucalyptus has also come for its use in social forestry because social forestry is always done on a massive scale, raising large plantations.

What are the views in favour of the use of eucalypts in social forestry, and what are the views against it?

Views in Favour of Eucalypts

According to Kaikini (1967) eucalyptus is highly suitable for social forestry. To meet the raw material demand of pulpwood industries which began to increase very fast during the 1960s, fast growing species were needed which could adopt themselves to the different agro-climatic zones existing in India. After several other trees were tried the genus eucalyptus was found to be ideal. Pryor (1976) and Tiwari (1983) has expressed similar opinion.

Davidson (1985) supports the introduction of eucalypts in social forestry programme, because of the tree's almost unique ability to adjust and adopt to a wide range of sites and at the same time grow more rapidly than other genera under widely varying environmental conditions.

CSE (1985 a) report that private tree-farmers and foresters love the tree and consider it a God-sent gift which grows well in dry conditions, does not get browsed, coppices well, fetches a high price and stabilises incomes.

Pro-eucalyptus environmentalist Karanth (CSE, 1985 a) states: *Claims such as that eucalyptus has caused a drought in Bengal, that these trees do not cast shadows and that thousands of livestock might go without a blade of grass to eat etc are all too fantastic to merit a rebuttal*... The Gujarat forest department, India, states: *Eucalyptus is an unmixed blessing for Gujarat farmers.*

CSE (1985 a) also report that in Haryana, India, nearly 40,000 hectares of crop land is under eucalyptus, especially in the rainfed areas of Ambala, Karnal and Kurukshetra. Some 150,000 hectares of eucalyptus plantations exist in Maharashtra, India, in combination with other species, of which some 25,000 hectares are irrigated. In Uttar Pradesh, India, eucalyptus covered 82,000 hectares by 1979. In Karnataka, India, about 1,33,000 hectares are under eucalyptus. Eucalyptus plantations are also seen increasingly in the states of West Bengal, India and Tamil Nadu, India.

Many farmers have become dye-hard eucalyptus supporters. Mishra (CSE, 1985 a) claims that *in Uttar Pradesh, India, we even insist that 10 per cent of plantations should comprise of species other than eucalyptus. But farmers in western Uttar Pradesh want only eucalyptus.* Shyamsundar (CSE, 1985 a) has stated *Last year some 5 million seedlings (of other species) were wasted in our nurseries due to lack of demand.*

Dinesh Kumar (CSE, 1985 e) argues that eucalyptus plantations have a definite role to play in agroforestry systems. Eucalyptus plantations in the midst of farmlands act as windbreaks, increase humidity and cut down radiation to increase photosynthetic activity in nearby agricultural crops. These advantages have resulted in 23% and 24% increase in the yields of wheat *(Triticum vulgare)* and mustard *(Brassica sp.)* respectively in Gujarat, India, and 40-43%, 39-47%, and 23-64% increases in yields of groundnuts *(Arachis hypogia)*, pulses *(Cajanus cajan)* and millets *(Panicum sp.)* respectively in Andhra Pradesh, India. In Gujarat, India, where irrigated farm forestry is progressing in a big way many farmers have started taking intercrops with eucalyptus. Patel (CSE, 1985 e) has been experimenting with the possibility of taking grain and pulse crops. He already grows two species of fodder grasses.

Khoshoo (CSE, 1985 f) is of the opinion that *eucalyptus can be a blessing in a captive plantation... eucalyptus plantation can be grown successfully on marginal or degraded land without any major disadvantage.*

Concerned with the increasing criticism against eucalypts. The Government of Karnataka, India, had set up an expert committee called Environmental Planning Committee of Karnataka (EPCK) (CSE, 1985 b). After extensive deliberations the committee submitted a report which states, *inter alia,* that *if eucalyptus plantations under social forestry replace agricultural crops in drought-prone areas to a limited extent, that would be ecologically beneficial.*

According to Chaudhury (1986), eucalypts are the most suitable tree for farm forestry because of their fast growth, thin crown, straight bole and easily coppicing nature.

Poore and Fries (1987) also consider eucalypts as 'ideal genera' for afforestation purposes, especially for the previously tree-less sites, because of their rapid growth rate and their ability to survive even on poor soils, in particular those deficient in phosphorous.

That eucalypts can grow well even on poor soil is subtantiated by Saxena (1992). He has reported that eucalyptus has been grown in areas of low productivity surrounding the metropolitan town of Bangalore, Karnataka, India where eucalyptus logs and poles are brought by the paper mills and by construction companies. He has emphasizes the fact that the area where eucalypts are grown is predominantly semi-arid with uncertain rainfall, and that eucalyptus has replaced an inferior food-grain, ragi *(Eleucine coracane)*.

According to Rao (1995) eucalyptus were planted to check desertification in the Nile valley, indicating that the tree has the ability to grow under very adverse, desert-like conditions.

Views Against

Kang (CSE, 1985 f) claims that many farmers have stopped growing eucalyptus along field boundaries because it depletes water on both sides of the boundaries. He had found that eucalyptus attracted crows which then damaged grain crops. Kang recommends that a legume like leucaena must be planted in between eucalyptus to preserve soil fertility. Bhumbla counters Kang's observation by saying (CSE, 1985 f) *it must be pretty bad agriculture to begin with which gives a 24 per cent increase in yield of wheat simply by planting eucalyptus along field boundaries.*

According to Krishnamurthy (CSE, 1985 a), foresters in Karnataka, India, who had felled natural forests in the Western Ghats for eucalyptus, now restrict its planting to lower rainfall zones of 500 mm to 1,125 mm. He also states *The depressing effect of drought on fields of rainfed crops (millets, pulses and oilseeds) coupled with the lack of remunerative price support for these crops, is driving the farmer to such frustration that opportunities offered by industrial interests for turning their lands into eucalyptus plantations are finding ready acceptance. And the forest department is encouraging the farmers to raise eucalyptus through the massive distribution of seedlings free of cost.*

Shiva and Bandyopadhyay (1987) strongly criticised the introduction of eucalypts in soical forestry. According to them the choice of eucalypts in social forestry is not apt *vis a vis* the needs of the common man whose priorities are fuel, fodder, green manure *etc*, rather than industrial products such as paper and rayon.

On Balance

The gist of arguments *for* the use of eucalypts in social forestry are:

(a) they are exceptionally productive and remunerative

(b) they do not have any different (or particularly adverse) impact on environment than other trees used in social forestry.

The critics of eucalypts question both these assertions and declare that eucalypts should not be used in social forestry or for any other purpose.

In the following sections we have examined in detail the arguments and counter-arguments on all specific aspects of eucalypts—beginning with productivity. We have put particular emphasis on the evidence of quantitative and controlled experiments. On balance we find that eventhough there is no *conclusive* evidence in literature, there is *stronger scientific* evidence *for* eucalypts than against it—indicating that eucalypts can be gainfully used in social forestry provided certain precautions are taken, mainly the proper choice of species and proper manner of harvesting (to avoid overkill and thus safeguard against all such negative impacts on water and soil that overexploitation of *any* tree may cause). Our studies based on controlled experiments performed in undisturbed *Euclayptus hybrid* plantations, described later in this book, have established this fact beyond a shadow of doubt.

These points would become progressively clearer on perusal of the following sections of this chapter.

Productivity of Eucalyptus

The most basic and powerful argument in favour of using eucalyptus in social forestry has been based on the belief that the genus eucalyptus is one of the fastest growing trees in the world or, more precisely, *one of the fastest growing trees which is also one of the easiest to grow.*

All those who favour eucalyptus emphasize this argument again and again but those who are against eucalyptus have been challenging this belief.

Evidence that Eucalypts is Exceptionally Productive

Khan (1959) studied plantations of eucalyptus *(Eucalyptus tereticornis-globulus)*, casuarina *(Casuarina equisetifolia),* and acacia *(Acacia auriculiformis)* raised in Sriharikota island, Andhra Pradesh, India. He acquired growth data from four-year plots and statistically analyzed the data. This revealed that eucalypts produced more woody biomass than the other species in identical period of time. Whereas eucalyptus plantation generated an average of 1.3 cubic feet of wood, acacia generated less than a fourth of it (0.34 cubic feet) and casuarina about 3% less (1 cubic feet).

Kawahara *et al.* (1981) compared the growth rate of *E. camaldulensis* with other fast growing trees and concluded that the productivity of *E. camaldulensis* (17-59 tonnes/ha year) puts it in favourable light in comparison to the productivity of other trees (15-30 tonnes/ha year).

In 1983 Chaturvedi reported findings of an experiment conducted at Kanpur, India, specifically to decide which of the six fast-growing trees—*Acacia auriculiformis, Albizzia lebbeck, Dalbergia sisso, Pongamya pinnata, Syzigium cumini and Eucalyptus hybrid* were suitable for promotion in the social forestry programme in India. The studies were conducted under controlled conditions so as to eliminate extraneous factors (such as grazing) from influencing the productivity data. Their findings, summarized in Table 1, support the observations of earlier (and later) workers about the high productivity of eucalyptus.

In a plantation raised on silty clay loam soil, Gurumurti and Rawat (1989) observed that in *Acacia nilotica, Prosopis juliflora, Acacia tortilis* and *Cassia siamia* the 'bole biomass' (the woody portion) constituted only 30-40% of the total biomass, in *Eucalyptus hybrid* and *Casuarina equisetifolia* the

Table 1. Productivity of six trees explored for social forestry (Chaturvedi, 1983); the data is for growth across one year span in trees of identical age

Sl. No.	*Species*	*Biomass Production (gms)*			
		Shoots	*Roots*	*Leaves*	*Total*
1.	*Acacia auriculiformis*	1023.5	361.6	327.9	1713.0
2.	*Albizza lebbeck*	1132.4	085.6	136.8	2354.8
3.	*Dalbergia sissoo*	1129.3	775.5	99.7	2004.5
4.	*Pongamia pinnata*	168.0	274.7	77.5	519.7
5.	*Syzigium cumini*	1278.0	593.7	514.3	2386.0
6.	*Eucalyptus hybrid*	2519.8	2094.3	594.9	5209.0

bole biomass, but formed as much as 80% of the rest of the tree. From these observations the authors concluded that eucalyptus and casuarina are amenable to close cropping and are ideally suited as pole crops. Their observations also indicate that eucalyptus (and casuarina) generate much more quantities of *profitably utilizable biomass* than the other trees studied.

Somachai *et al.* (1990), recording the performance of eucalyptus in Japan, have observed that the tree yielded about 44.9 tonnes/ha year of biomass which was over 30% more than the yield from *Acacia mangium* plantation of identical age.

Bargali and Singh (1991) report that the net primary productivity of a eucalyptus plantation (23.4 tonnes/ha year) was similar to that of the *Populus deltoides* plantation (25 tonnes/ha year) and a natural sal forest (22 tonnes/ha year). However the net nutrient uptake of eucalyptus was lower than that of the populus plantation and the natural sal forest. This, the authors believe, gives eucalyptus a net advantage *vis a vis* productivity—soil nutrient ratio.

In a recent report from Neyveli Lignite Corporation (NLC), Tamil Nadu, India (Shankran, unpublished internal report, 1995), productivity of *Eucalyptus tereticornis* has been compared with *Acacia auriculiformis* and *Leucaena leucocephala* grown within the NLC campus (Table 2). The study clearly reveals that *E. tereticornis* is by far the most productive of the three fast-growing trees studied.

Table 2. **Productivity (Timber wt.) data of different tree species at Neyveli Lignite Corporation (field data)**

Sl. No.	*Tree species*	*Productivity*	*Rotation (years)*
1.	*Eucalyptus tereticornis*	30-60kg/tree	5-7 years
2.	*Acacia auriculiformis*	25-30kg/tree	7 years
3.	*Leucaena leucocephala*	20-25kg/tree	6 years

The Counter Arguments

Shiva and Bandyopadhyay (1987) have stated, '... *having found that all recognised and established scientific information is illegetimising eucalyptus as a fast growing tree, the forest establishment of the country stopped their dependence on these data and initiated their controlled experiment immediately after the domination of eucalyptus in the social foresty programmes got challenged in 1981'*. Their allusion is to the data reported by Chaturvedi (1983), summarized in Table 1, which they (Shiva and Bandyopadhyay) believe to be the only study of its type ever conducted. They have then gone on to blast the so called 'Kanpur experiment' as the *'only straw* (available to the proponants of eucalyptus) *to hold on to...'*

Shiva and Bandyopadhyay (1987) have also quoted from Bhumla (1984) to say that there is no data available to prove that eucalypts produces more biomass than other indigenous trees. They have also quoted from Quereshi (1967) and have mentioned *'a comparison of growth rate of ten species by the Gujarat Forest Department'* to say that eucalyptus is not one of the fastest growing but among the slowest of the faster growing trees.

On Balance

Eventhough Shiva and Bandyopadhyay have used strong words to denounce eucalyptus, they give no hard data at all on any experiment done *simultaneously on eucalyptus and other species* to substantiate their arguments. Nor have we found in literature any such authentic data. On the other hand numerous authors from India as well as abroad have reported through independent experiments conducted under different regions and across a time span of several decades- summarized by us in the proceeding section- which indicate that eucalyptus is indeed one of the most productive trees and *generally* yields more valuable biomass than such other fast-growing trees as acacia, casuarina, leucaena, and cassia.

Eucalyptus and Surface Run-off

It has been alleged that the surface run-off is very high in eucalyptus plantations, because of scarce undergrowth and other adverse edaphic conditions created by the tree. It has also been alleged that this high surface run-off leads to soil nutrient deficit, besides adversely affecting groundwater recharge.

Views in Favour of Eucalypts

Chinnamani *et al.* (1965) have observed that the proportion of run-off from the eucalyptus and acacia was similar to that from the shola forests. Similar results based on a three-year study have been reported by Samraj (1977).

According to Mathur *et al.* (1980), the presence of excellent undergrowth which is generally allowed by an open canopy of eucalyptus *actually reduces* run-off, especially during rainy season. They have reported results of their experiments in support of this observation.

Poore and Fries (1987) have stated that eucalypts can be, and has been, used for erosion control. They suggest that even if surface run-off is noticed, it can be alleviated by terracing—a practice which is also beneficial for the establishment and growth of eucalypts on steep dry sites. Earlier de le Lama (1984) had also suggested terracing to control soil erosion.

Views Against

Stein (1952) has reported that in steep dry areas where *E. globulus* has been planted, understorey development and litter build-up were insufficient to prevent surface run-off. He believes that eventhough *Eucalyptus globulus* is a fast growing, heavy crowned tree which casts a dense shade, it gives little litter. This makes eucalyptus plantation susceptible for rapid surface run-off. He also had reported that in Southern India, where rainfall is less than 750 mm the failure of an understorey to develop coupled with a weekly developed forest floor, leaves the soil exposed to runoff under such condition.

Mathur *et al.* (1976) and Balagopalan (1986) also have reported that eucalyptus plantations hasten soil loss.

According to Poore and Fries (1987), eventhough the presence of litter and ground vegetation greatly controls the amount of run-off, this would certainly vary according to the climate. The ground vegetation of eucalyptus forest is sparse in dry climates due to root competition and perhaps, allelopathic effects. But the situation would be different in less drier climates.

On Balance

It appears that whether eucalyptus leads to an increase in water run-off and soil loss depends upon what it is replacing. If *Eucalyptus hybrid* is replacing a natural stratified forest then the water run-off and soil loss will be *relatively* increased. If, on the other hand, *Eucalyptus hybrid* is planted in agricultural land or degraded land devoid of tree cover—as envisaged in the government plans—then the water run-off and soil loss will decrease.

The intensity of surface run-off also greatly depends upon the type of soil as well as the species planted. It has been reported that in some species of eucalyptus the root-reticulation pattern is such that it binds the soil particles together, thereby the surface soil escapes from being eroded. Much also depends upon the rainfall regime of the area. If the intensity of rainfall is high the surface run-off also will be at the maximum. This condition can be noticed in almost all types of plantations. Hence selection of proper regime and right species would greatly control the soil loss in eucalyptus plantation. Selection of right species would also lead to an improvement in the undergrowth species development, which would also greatly control the surface run-off. The experiments of these authors, described later in this book, have provided strong evidence to support this contention. We have found that the diversity of undergrowth in the plantation of *Eucalyptus hybrid* is as good as in other monocultures or mixed cultures. There is thus as good protection against run-off and soil erosion in eucalyptus plantations (if proper species are used) as provided by other tree species.

Eucalyptus and Water Uptake

The most often-repeated allegation against eucalyptus is that it consumes large amounts of water thereby reducing soil moisture as also lowering the groundwater table.

Views in Favour of Eucalypts

According to Shyamsundar (1983 a) the criticism against eucalyptus that it lowers the ground water table is baseless as the roots of the eucalypts rarely go lower than 3-4 metres, hence it could not tap subterranian water.

Lima and O'Loughlin (1984) support this view. They strongly oppose the statement that the shallow root system of eucalypts is the main cause for depleting the soil moisture. They are of the view that the lateral spreading and depth of penetration of the root system vary with species and this has to do with intensity of water uptake. Hence selection of right species will prevent the problem of high water uptake. They also report that the overall soil water regime of eucalyptus forests do not differ from that observed in pine plantations.

According to Foley and Bernard (1984) eucalypts roots can break up the soil structure or even a subterranean layer of impervious hard pan. Hence eucalypts can improve rain water percolation, creating a net positive effect on the ground water level.

Dinesh Kumar (1984) states that several trials in Australia have proved that eucalyptus is the most efficient utilizer of scarce water resources. He concludes that the species itself is a good drought resistant one.

As far as the possibility of eucalyptus plantations lowering the ground water table is concerned, the Economic and Planning Council of Karnataka (EPCK) report (CSE, 1985 b) argues that the roots of *Eucalyptus hybrid* rarely go lower than 3 metres to 4 metres deep and usually do not spread out (laterally) more than 1.5 metres. This means that *Eucalyptus hybrid* only consumes subsurface seepage water, and it cannot tap subterranean groundwater. This being so it is unlikely that *Eucalyptus hybrid* is responsible for wells running dry. The alter phenomenon can be due to several other factors such as: enormous increase in the number of irrigation pumpsets; continuous drought for consecutive years; and very high density of tree planting. In Davanahalli and Hoskote taluks of Karnataka state, India, for instance the number of energised irrigation wells have doubled within six years. Similar intensity of exploiting underground water is visible in several part of Kolar district, Karnataka, India. Chaturvedi (CSE, 1985 b) argues that the Terai region of Uttar Pradesh, India, where eucalyptus plantations have allegedly lowered the water table, is actually full of tubewells which might be the real culprits.

Ray (CSE, 1985 c) states that *it is a fact that streams have dried up in Nilgiris after the removal of the original shola forests, but it is uncharitable to blame it on eucalyptus. The eucalyptus plantations would have helped in recuperating the subsoil water had the leaf litter been allowed to remain on the ground and get converted into humus. But all the leaves are removed for distilling eucalyptus oil. In the plains, the local people remove the leaves and twigs for cooking purposes. In such a situation the soil is not in a position to absorb water and recuperate the water table.*

CSE (1985 c) has mentioned that the foresters repeatedly cite a 1972 study from the Nilgiris which reports that the annual transpiration of water in a *Eucalyptus globulus* plantation corresponds to a rainfall of 34.75 cm whereas potato fields use 65 cm of rainfall. The total rainfall in the area was 130 cm thus making available the balance 95.25 cm to cover interception loss, surface run-off, evaporation, deep percolation, water yield and soil moisture storage.

Mathus (CSE, 1985 c) believes that *it is agriculture crops which are responsible for depletion of water resources and not eucalyptus plantations. Water use for some of the agricultural crops like wheat (Triticum vulgare), paddy (Oryza sativa), sugarcane (Saccharum sp.) and millet (Panicum sp.) is 38 cm, 104 cm, 163 cm and 64 cm respectively.*

Patel (CSE, 1985 c), one of the first eucalyptus farmer of Gujarat, India, switched over to eucalyptus because there was not enough water to grow cotton, which is a heavy user of water. Now that he has been farming eucalyptus, he claims, the water table in the region has stabilised. Patel also claims that the plantation has improved the water absorption capacity of the land as compared to neighbouring farms because eucalyptus roots break the soil and thus increase the seepage. During one night in 1976, water accumulated from 12 inches of rainfall, was absorbed within one hour on Patel's eucalyptus farm. On the contrary, the water on some of the neighbouring farms stood for three to four days after the rain.

Poore and Fries (1987) have quoted that drawing of soil moisture depends on stand density, and soil and environmental conditions. They have quoted from Lima and O'Loughlin (1984) and have summarised their findings to say that in alpine dry sclerophyll condition, soil water regime does not differ between eucalyptus forest, grassland, and herb field.

Poore and Fries (1987) have also reported that the effect of eucalyptus in reducing water yield is probably less than that of pine and greater than that of other broad-leaved species; but all species of trees reduce water yield compared with scrub or grass. The yield of eucalyptus timber (11.1 m^3 mean annual increment) by far offsets the value of that part of the water losses that would have been added to the ground water. They are of the opinion that though the water-yield was reduced by about 20% compared with that from open ground, it is probable that a somewhat similar loss would have occurred under any other tree crop. They also have reported that the overall soil water regime of eucalyptus forest does not differ from that observed in pine plantations which is also a fast-growing tree.

According to Srivastav (1993), eucalyptus has high water holding capacity in the soil. According to his study conducted in the arid region in Surat District of Gujarat state, India, there was more soil moisture under eucalyptus than a nearby open area even after three consecutive drought years!

According to Rao (1995), the argument that eucalyptus absorbed as much as 60 gallons of water a day is not based on reality. For this, plantation area with eucalyptus planted two metres apart should receive nine cms of rain per day. Taking half of this as average, the yearly uptake of water should be 1642.5 cm which is unbelievably high. But no part in India where eucalyptus has been planted receives so much of rain. In fact eucalyptus grown in Nilgiri, Tamil Nadu, India, for over a century uses only 35 cms of the 135 cms of rainfall there.

Rao (1995) has also stated that the *Eucalyptus hybrid* adopted in the Indian subcontinent is not a wasteful consumer of water. But, is on the contrary, one of the most efficient utilizers of scarce water, producing more timber for water consumed than many other native species. He also states that scientists in Brazil (which have more than a million hectares of land under eucalyptus) confirm that eucalyptus plantations consume less water than the same area under natural forests.

Views Against

Stein (1952) has stated that in closed plantations eucalyptus has a great water demand. This together with an extensive and dense root system enables it to compete successfully for available soil moisture, especially with smaller, shallow, rooted plants. Thus eucalyptus uses all the water available to the soil.

According to Heith and Karschon (1967), a study conducted in the central coastal plains of Israel (rainfall 600 mm, a dry period of 3 to 5 months) where eucalyptus plantation was compared with an open ground, showed that eucalyptus made use of all the water available to it.

Maheshwata Devi (1983) and Bahuguna (1984) also believe that eucalyptus consumes more water than other trees. According to them the streams feeding agricultural lands in the vicinity of eucalyptus plantations have gone dry.

They also claim that in arid regions the high water uptake by eucalyptus interferes with processes which replenish soil moisture and recharge ground water leading to soil aridisation and ground water depletion.

Chaturvedi (CSE, 1985 b) says that *in any are the same number of eucalyptus trees will consume more water than any other species during the same period.*

Gupta (CSE, 1985 b) points out that in low rainfall areas, eucalypts roots form a dense network just below the soil surface to extract every bit of moisture.

The Economic and Planning Council of Karnataka (EPCK) report (CSE, 1985 b) agrees that eucalyptus plantations are heavy consumers of water on a per hectare basis. But, it argues that *one has to look at the total water balance including the impact of transpiration losses and the extent of percolation into the soil.*

The statement that eucalyptus is drought resistant is strongly questioned by Shiva and Bandyopadhyay (1987). According to them the shallow root system of eucalypts prevents it from surviving through permanent water scarcity, unlike the indigeneous species that have been proved to be genuinely drought resistant. They also have mentioned a study conducted by the hydrological division of the CSIRO in Australia on the hydrological impact of eucalyptus on water resources to say that the efficiency of utilizing water by eucalyptus is greatly controlled by the rain-fall regime of the area. During years with precipitation less than 100 mms, deficits in soil moisture and ground water were created by eucalyptus. A permanent water deficit was avoided by significantly high rain fall of 1477 mms in one of the 5 years studied. Table 3 summarizes the results of the long term hydrological study showing that when rain fall is of the order of 1000 mms or less, eucalyptus plantations create deficits both in the soil moisture and ground water.

The accusation of Shiva and Bandyopadhyay (1987) against eucalyptus is that its fast growth requires excessive water and its lateral roots are spread in such in a way that

Table 3. **Changes in soil moisture and ground water in eucalyptus catchment (mms)**

Sl. No.	*Year*	*Precipitation*	*Soil Moisture*	*Ground water*
1.	1975	1477	+29	+27
2.	1975	914	-87	-14
3.	1976	883	-49	-33
4.	1977	983	+49	-12
5.	1978	900	+30	-19

recharge of water through percolation becomes impossible. Hence, ground water does not get recharged. Also the vast network of root system just below the soil surface extracts every bit of moisture made available to the soil by precipitation. They also question the statement that the roots of eucalypts rarely go lower than 3-4 metres and that it cannot tap subterranian water. According to them the talk of the tapping of the underground water resources through the tap root as the only process of depleting ground water by the trees is either sheer illiteracy of elementary arithmetic and biology or it is a calculated attempt in misinforming the lay public and non-specialists in the decision-making bodies. Shyamsunder (1983 b) also have expressed similar view. Little evidence has been found of the water consumption of eucalyptus under natural conditions of unlimited water during the summer months nor data comparing eucalyptus with other tree species under such conditions.

On Balance

The evidence discussed above has sought to answer two key questions: do eucalypts use more water or have a greater effect on the water regime than other trees and are eucalypts more efficient in their use of water than other trees?

One conclusion can be clearly drawn, as pointed out by Poore and Fries (1987), that depending upon circumstances, eucalyptus has been used from time to time to lower water-tables in swampy areas either to dry out the soils or to control mosquitoes. Here, the effects clearly accomplish their purpose and are beneficial. If, however, eucalyptus lead to the reduction in volume of an aquifer which is used downstream for domestic water supply or for irrigation water, the effects are likely to be considered harmful. In all such cases it is important to consider the purpose of planting (fuelwood, shade, shelter, poles *etc*), the various uses that might be made of the water, and the total benefits and costs in the local socio-economic context.

According to Foley and Bernard (1984), whether a eucalyptus plantation will affect the water-table depends greatly on the hydrological and physical properties of the soil. It is also determined by what kind of vegetation it replaces. If the previous crop was a water hungry one, the water-table may well rise. If the eucalyptus are being planted to replace slow-growing scrub, on the other hand, in an area with a sensitive hydrology, it is quite possible that the water-table might fall. The effects can only be predicted through a careful site survey. Further, as in the case of other aspects of eucalyptus controversy, the strong views for or against eucalypts invariably emanate from piece-meal observations made without due consideration of the context or the setting under which certain positive or negative impacts of eucalypts were observed. There is also the recurring theme of blaming the entire genus of eucalyptus for the adverse impact of one or two of its species.

Our experiments, reported later in this book, have established that *Eucalyptus hybrid* plantations do not deplete soil moisture and their performance in this report always compared favourably with plantations of other tree species.

Eucalyptus and Rate of Transpiration

It has been alleged that transpiration losses—loss of water from the surface of the leaves—is very high in eucalyptus tree genus. It has been said that this is due to the fact that eucalypts do not have the mechanism to control transpiration.

Views in Favour of Eucalypts

Dinesh Kumar (1984) has refuted the allegation that eucalyptus has a high transpiration rate. According to him eucalyptus being a xerophyte has a low transpiration rate and it controls stomatal openings according to water availability without serious reduction in biomass production. Similar findings have been reported by Brown *et al.* (1976), Ackerson (1980) and Singh *et al.* (1993).

According to Poore and Fries (1987), majority of eucalyptus species do have some control over the rate of transpiration, which helps them to survive drought stress during some part of most years, and which is apparently related to the rainfall regimes of their natural habitats. They also have reported that average annual evapotranspiration in pine plantations is in the same order of magnitude as that observed in eucalyptus forests.

Views Against

Pryor (1976) has reported that the transpiration rate of eucalyptus remain high even when the water supply from the soil has dwindled.

High transpiration rate by eucalyptus has also been alleged by Shiva and Bandyopadhyay (1987). They have listed in Table 4 the transpiration rate of eucalyptus and other tree species to show that eucalyptus shows high transpiration rate compared to other such trees.

Table 4. Transpiration of eucalyptus and other tree species (in mm)

Sl. No.	*Species*	*Transpiration*
1.	Eucalyptus	1200
2.	Eucalyptus	1248
3.	Eucalyptus	1136
4.	Eucalyptus	5526
5.	Eucalyptus	1255
6.	Mixed Forest	140
7.	Mountain Rain Forest	870
8.	Birch	564
9.	Beech	456
10.	Spruce	516
11.	Pine	282

On Balance

As in other aspects of eucalyptus controversy, the stress has been on the 'black' or 'white' areas with little regard to the 'grey' areas in between. There is as much hard data suggesting that eucalyptus causes heavy transpiration losses as there is evidence that eucalyptus does not. Several species of eucalyptus have the ability to adjust to different ranges of habitats. If eucalypts are grown in an area where there is surplus ground water, they would make use of the water for their growth. At the same time if they are grown on moisture-lean soils, they adopt themselves to that habitat. This can be possible only if they have some mechanism to control their water usage, including rate of transpiration. Or else it would be impossible for them to survive in drought-like conditions.

In the present study it has been established that the *Eucalyptus hybrid* plantations were not wasteful utilizers of water resources and their effects on the soil moisture were always comparable with other tree plantations advocated as being ecologically superior to eucalyptus by a section of environmentalists.

Eucalyptus and Soil Nutrients

A number of possible effects of planting eucalyptus on the nutrient balance have been suggested. One of the criticisms against eucalypts is that they may deplete the nutrients of the site, particularly if they are grown and cropped for several rotations.

Views in Favour of Eucalypts

Attwill (1966) in his study under mature *E. obliqua* forest in the great dividing range of southeastern Australia reports that eucalyptus does enrich soil nutrients and the principal contribution is made by leaching nutrients from the leaves.

George (1979) has estimated the nutrients contained in the rainfall, stemflow and throughfall in a plantation of *Eucalyptus hybrid* at Dehra Dun, India. The data on concentration of nutrients is presented in Table 5 and total yields (in kg/ha year) in Table 6. It is seen that concentration of salts in stemflow and to a lesser extent in throughfall are greater than in the rainwater. However, it is unclear whether these additional amounts come from leaching of the foliage or by washing aerosols and dust from the leaves.

Table 5. Concentration of nutrients in stemflow, throughfall and rainwater

Sl. No.		*Nutrients (ppm)*				
		K	*Ca*	*Mg*	*N*	*P*
1.	Stemflow	3.07	3.01	0.19	0.18	0.17
2.	Throughfall	0.70	0.65	0.15	0.15	0.01
3.	Rainwater	0.31	0.35	0.15	0.10	0.01

Table 6. Nutrient return through stemflow, throughfall and rainwater (kg/ha year)

Sl. No.		*Nutrients*				
		K	*Ca*	*Mg*	*N*	*P*
1.	Stemflow	3.9	3.8	0.2	0.2	0.1
2.	Throughfall	9.4	8.8	2.0	2.0	0.1
	Total	13.3	12.6	2.2	2.2	0.2
3.	Rain water	5.2	5.9	2.5	1.7	0.2
	Grand Total	18.5	18.5	4.7	3.9	0.4

Raison and Crane (1981), experimentally found greater rates of phosphorus removal when harvesting *Pinus radiata* compared with *E. delegatensis* (Table 7).

A study on the properties of soil under eucalyptus plantations in Kerala by Alexander *et al.* (1981) indicated high fertility in uncoppiced as well as first coppiced plantations.

Jha (1984) reports that eucalyptus enriches the soil through the leaf and twig litter.

Lima and O'Loughlin (1984) have presented an interesting evidence of the interaction of rainfall with the forest canopy. They compared the nutrient contents of rainfall, through directfall and stemflow in four different kinds of eucalyptus forest. It was seen that there is a consistent

Table 7. Quantities of rates of phosphorous exported from the forest when harvesting *E. delegatensis* and *P. radiata* on short and long rotation

Sl. No.	*Parameter*	*Eucalyptus (E. delegatensis)*		*Pine (P. radiata)*	
1.	Tree rotation (yr)	18	57	18	40#
2.	P exported (kg p/ha)				
	in stemwood	9	17	28	56
	in bark	4	8	18	24
	in bole	13	25	46	80
3.	Rates of P harvested in boles				
	as per wood (q pt/Wood)	97	51	258	169
	as per time (kg p/ha.yr)	0.73	0.44	2.53	1.97

#includes 4 commercial thinning at ages 16,22,28, and 34 prior to clearfelling at age 40.

enrichment of rain water after it passed through the canopy, especially in terms of sodium and potassium. The leached sodium was about two times as much as is found in the litter fall and potassium 1-3 times (Table 8). These experiments indicates that in an eucalyptus plantation soil nutrient enrichment also occurs through the rain water intercepted by eucalyptus trees. There has been no report contradictory to the findings of Lima and O'Loughlin.

According to Foley and Bernard (1984) the allegation that eucalypts extracts all the nutrients available in the soil is not just. They are of the view that all plants extract nutrients from the soil. When the plants are harvested and biomass is removed, the soil nutrients are bound to be lost from the plantation. This could happen with any tree species. The rate of loss of nutrients is dictated by the rate at which biomass is removed. If under such circumstances a high yielding tree genus like eucalyptus reduces soil nutrients, so will other plants like rice, sugarcane, leucaena, prosopis or any other fast growing tree.

Foley and Bernard (1984) report that the rate at which the depletion of soil nutrients occur, and the nature of the

Table 8. Leaching of nutrients from eucalyptus forests canopy by rainfall

Sl. No.	*Species*	*Process*	*Kg/ha year*					
			K	*Ca*	*Mg*	*Na*	SO_4	*P*
1.	*E. signata-E. umbra*	P	3.4	3.2	5.9	5.0	9.6	-
		T	8.5	14.0	7.2	44.0	17.0	-
		S	0.9	0.8	1.1	8.1	0.7	-
2.	*E. nilanophloia*	P	2.6	1.9	0.7	3.7	-	0.1
		T	22.1	9.3	8.1	6.4	-	0.5
		S	0.7	0.5	0.5	0.2	-	0.01
3.	*E. obliqua*	P	4.2	1.3	1.4	17.9	-	-
		T	15.4	6.3	6.0	37.2	-	-
4.	*E. obliqua*	P	2.0	2.7	5.4	16.8	-	-
		T	13.4	8.0	7.3	25.4	-	-

P= Rainfall

T= Throughfall

S= Stemflow

problem that arise from it, depends to a large extent on the type of soil and its physical structure rather than the tree species. Soils which are inherently fertile can support high yielding crops for many years without the productivity declining. Others may be depleted very quickly. But the case against eucalyptus is no different from that which might be made against any other crop.

The report of the expert committee, set up by Government of Karnataka, India, (EPCK), to examine the ecological impacts of eucalyptus (CSE, 1985 d) has noted that *the question of whether Eucalyptus hybrid enriches the soil also depends on what Eucalyptus hybrid replaces and on the planting density. If Eucalyptus hybrid replaces a tropical rain forest, then it will return less nutrients to the soil than the rain forest. But eucalyptus will enrich the soil in the case of marginal agricultural land or unwooded or degraded areas.*

The report has further noted that the impact of *Eucalyptus hybrid* on the soil depends on the planting density. It appears that if the planting density increases above 10,000 trees per hectare, nitrogen and phosphorous deficiency may result. In the state of Karnataka, India, the planting density is about 3,000 trees per hectare; at these low densities, there is no evidence that eucalypts depletes the soil. Indeed, by withdrawing the nutrients from the lower levels in the soil and depositing them on the surface, the nutrient status of the top soil is enhanced in eucalyptus plantations. To sum up, the report says, although natural forests should not be replaced with *Eucalyptus hybrid* monoculture, there are no compelling and perceived grounds for discouraging eucalyptus cultivation in marginal agricultural or degraded lands, from the viewpoint of nutritional adequacy. The report stresses the fact that if proper management practices in eucalyptus plantations are not adopted, quick rotations of eucalyptus may definitely affect the nutrient levels in the soil.

Chaturvedi (CSE, 1985 d) states that there are genuine cases of 'second rotation decline'. He believes that eucalypts grow fast and are generally harvested at short rotations in commercial farming. Their demand for moisture and nutrients cannot keep pace with felling in such cases. It may, therefore, by necessary to rotate eucalyptus with other species. These species should be nitrogen fixing, shallow-rooted and deciduous. Research for the choice of such species and their cycle of planting is needed.

Patel (CSE, 1985 b) argues that eucalyptus leaf litter builds up rapidly in the soil. He also emphasizes that, because of little sunlight that reaches, and heavy shade, it also decays very quickly, thus enriching the soil.

Kushalappa (1984) has reported an improvement in organic carbon, phosphorus, and potassium contents in the soil in eucalyptus plantations thus providing a strong evidence that eucalyptus monocultures are not detrimental to soil fertility.

Poore and Fries (1987) believe that eucalypts may improve soil characteristics when planted on degraded or deforested sites by improving the structure of the surface soil, by penetrating relatively impermeable layers of sub-soil, and by drawing up nutrients from depth. These authors have also quoted from Singhal *et al.* (1975) and Singhal (1984) mentioning experimental results to show that humification is faster under eucalyptus. They also report that because of this reason the chances of loss of organic matter from the soil is also reduced. According to them the humification is brought about by eucalyptus debris, which produce humic acid, increase the polysaccharides, and hastens the rate of polymerisation with reduction in dispersion, thus enhancing the fertility of the soil. Jamet (1975) and Jha and Pande (1984) have stated similar views.

The natural eucalyptus forest appears to control the leaching and run-off of nutrients slightly better than other natural forests (Poore and Fries, 1987). The effects of evercropped eucalyptus on soil depends upon the state of the soil in which the trees are planted; beneficial in degraded sites, probably not so when replacing indigeneous forests. When eucalypts are planted in bare sites, there is an accumulation and incorporation of organic matter.

In a recent study Pal and Ratwri (1991) have shown that even as biomass is removed from a eucalyptus plantation, no significant nutrient deficit is created. They explain that generally, the utilizable biomass of a large eucalyptus tree genus has larger proportion of wood, and comparatively lesser bark and branches. Since bark and branches generally store higher amount of nutrients than rest of the tree, more nutrients are lost when trees with larger proportion of bark and branches are harvested than the ones, like eucalypts, of smaller ones.

George and Varghese (1991) report that only certain species of eucalyptus are high extractors of soil nutrients and that entire genus should not be blamed on this count. Their experiments reveal that *E. globulus* takes up 184 kg/ha year of NPK and returns as much as 102 kg/ha year to the soil while *E. hybrid* takes up 113 kg/ha year of NPK and returns only 31 kg/ha year to the soil. In other words only certain and not all species of eucalyptus cause nutrient stress on the soil.

In Zimbabwe, Sanginga *et al.* (1991) found no evidence that eucalyptus is nutritionally more demanding than *Leucaenia leucocephala* and *Casuarina cummighamiana.*

Views Against

Pryor (1976) has observed that in Australia eucalypts are seen to thrive only on soils which are moderately deficient in nutrients. In highly deficient soils, according to Pryor, eucalypts do not grow.

Balagopalan (1986) states that eucalyptus plantation decreases organic carbon and other soil nutrients thus affecting the chemical and physical properties of soil.

Shiva and Bandyopadhyay (1987) claim that since eucalyptus genus is a fast growing tree, its nutrient requirements are excessively high and create a nutrient deficit since eucalypts returns only a small quantity of nutrient through its litter compared to its high uptake. They have quoted from Singh (1984) to buttress their argument (Table 9). Shiva and Bandyopadhyay (1987) also claim that eucalypts when planted on fertile agricultural land and harvested at short rotations, creates heavy nutrient deficits, destroying conditions of biological productivity.

Table 9. Nutrient deficit created by *Eucalyptus hybrid* plantations (kg/ha year)

	Nitrogen	*Phosphorous*	*Calcium*
Eucalyptus requires	217	100	1594
Eucalyptus returns	35	14	335
Annual nutrient deficit	182	86	1260
Deficient after 2nd rotation (20 yrs)	3640	1720	25200

Bahugana *et al.* (CSE, 1985 b) is of the opinion that leaves of eucalypts that fall to the ground do not retain any moisture; the leaves group themselves into heaps and remain in a dry state whatever be the intensity of rainfall. There is no mulching; and there are no dung droppings of any sort. As a result there is no decomposition of organic matter, vegetable or animal, no absorption of nutrients by the soils and no humus formation.

On Balance

It can be said that the nutrient uptake is a universal

occurrence that can be made about all fast growing tree crops (Prasad *et al.,* 1985). Moreover there is also no scientific evidence that eucalyptus have any special demands in this respect.

There are a large number of reports, augmented by hard data, from workers studying eucalypts across the world that supports the view that eucalyptus is no different than other fast-growing trees, perhaps better than many, in its impact on soil nutrients. In comparison there are lesser number of reports that accuse eucalypts of depleting soil nutrients and several of such reports are based on hastily arrived conclusions based on fleeting observations rather than rigorous and controlled experiments.

We feel that any fast-growing crop if harvested repeatedly in short rotations would deplete soil nutrients and eucalypts are not any worse than other trees in this respect. Frequent disturbance in the plantations such as brooming of leaf litter may substantially increase the potential for associated loss of organic matter and nutrients from the plantation.

The studies presented in this book have provided a firm footing to the arguments mentioned above. As the experiments conducted by us cover a large number of undisturbed plantations, studied over a large time-span, our findings can be taken as evidence sufficiently strong enough to settle this argument.

Allelopathic Effects of Eucalyptus and Undergrowth

Allelopathy is the deleterious effect of one plant on another through the production of chemical retardants that escape into the environment. The allelochemical and toxic effects of eucalyptus have been scientifically recorded and studied both in India and abroad. The report of Rice (1979), implicating *E. camaldulensis, E. baxteri* and *E. globulus* embodies one such study.

Views in Favour of Eucalypts

Mathur *et al.* (1980) compared plantations of *E. camaldulensis* and *E. grandis* with sal forest and secondary brushwood resulting 14 years after clearing of sal forest. Both the number of species and the vegetation cover were greater in the eucalyptus plantations and lesser in the sal forest. So were the amounts of litter, above ground phytomass, and below ground phytomass. The crown canopy (a measure of shading) were recorded as: eucalyptus 74.7%, brushwood 53.79% and sal 36.29%. These suggest better and richer vegetation under the eucalyptus. Further comparisons by Mathur and Soni (1983) and Rajvanshi *et al.* (1983) between sal and eucalyptus have led to similar conclusions.

According to the Forest Department, Government of Karnataka, India (CSE, 1985 d), *eucalyptus does not prevent undergrowth. Where rainfall and soil conditions are better, profuse regeneration of native trees, shrubs and grasses can be seen inside eucalyptus plantations. The lack of undergrowth in the arid areas is on account of already degraded soil, overgrazing, and removal of the leaf litter by the villagers.*

According to Rajan (CSE, 1985 d), the lateral roots of *Eucalyptus hybrid* ramify and occupy the top soil up to about 20 cm from the ground surface. Certain grasses grow well under eucalyptus plantations.

A study by the Central Arid Zone Research Institute at Jodhpur (CSE, 1985 g) points out that eucalyptus plants exploit subsurface moisture for their growth, leaving the surface moisture for undergrowth or other short duration crops. Another study reports that in areas which had only a coarse grass cover prior to planting of eucalyptus, the ground cover gradually changed to evergreen species.

According to Evens (1982), not all species of eucalyptus suppress other vegetation; under some species of 'lightly crowned eucalyptus', much undergrowth can be seen. It all depends upon the selection of right species of eucalyptus. This view is also supported by the present work.

Poore and Fries (1987) point out that the effects of eucalyptus on ground vegetation depend very much upon climate, this is mostly because of competition for water. The effects of reduced light created by eucalyptus trees are probably less than that are caused by other broad leaved trees like pines. Ground vegetation is less affected in wet conditions than in dry.

Williams (1990) argues that the seedling mortality which is reported to be high in eucalyptus plantation is not caused by the litter of eucalyptus alone. He is of the opinion that apart from competition, grazing and climatic condition are the prime factors responsible for seedling mortality. Hence, he feels, putting the blame solely on eucalyptus for its allelopathic effect is not fair. But if eucalyptus is grown in degraded and unwooded lands, undergrowth can occur provided the planting densities are low and of the order of a few thousand per hectare. Typically, there are other reasons for the absence of undergrowth such as excessive grazing, fires, fuelwood collection and soil erosion.

Vies Against

Story (1967), reported that some chemical exudates produced by eucalyptus were probably responsible for the allelopathic effect.

Del Moral and Muller (1969 and 1970) have reported allelopathic effect in eucalyptus plantation resulting in the absence of undergrowth.

Al. Mousawi and Al. Naib (1975) have also reported paucity of herbaceous plants in eucalyptus plantations in Iraq.

In India Swami Rao *et al.* (1984) were one of the first to report the allelopathic effect of eucalyptus. They found volatile and water soluble growth inhibitors in the leaf litter which did not encourage the germination and growth of other plants in the vicinity of eucalyptus.

Eucalyptus has been alleged to discourage undergrowth for reasons other than allelopathy. For example, according to Gupta (CSE, 1985 d) the heavy demand for water by eucalyptus prevents undergrowth.

Rajan (CSE, 1985 e) points out that since lateral roots of *Eucalyptus hybrid* and their ramifications occupy a large portion of the top layer of soil, leucaena remains stunted if grown near old *Eucalyptus hybrid* trees.

Shiva and Bandyopadhyay (1987) state that in arid regions, eucalypts inhibits the germination and growth of other plants through allelopathy thus posing a threat to food production.

Wilson and Zammit (1992), based on their experimental study at Brindabella Range, 30 km West of Canberra, Australia, report that eucalyptus litter limits germination and has the potential to limit distribution of understorey vegetation.

Rao (1995) has quoted from Bahuguna to say that planting eucalyptus in Uttarakhand area in the foot-hills of Himalayas has been mainly responsible for the stunting of fodder growth. According to Rao (1995) not even grass grows near eucalyptus.

On Balance

All-in-all, we conclude, there is no strong evidence that all species of eucalyptus have allelopathic effect; indeed most do not have.

From the foregoing discussion it is obvious that eventhough certain species of eucalyptus has been reported to produce toxins inhibiting the growth of other vegetation in the vicinity, *not all eucalyptus species are known to produce toxins.*

There is another phenomena which may create an illusion of allelopathy. This is absence of certain undergrowth species around a supposedly allelopathic tree *not because* of the repulsion caused by the tree but doe to the soil type and other ecological factors which, rather than the tree, may disfavour the presence of certain species of vegetation. Differences in soil type correspond to distinct floristic and structural formations within the forest or woodland. The understorey communities are correlated with soil properties as well. The casual factor in these relationships is unclear and requires experimental investigations.

The findings described in this book establish the fact that if a species of eucalyptus, which does not have allelopathic effect (such as *Eucalyptus hybrid*) is chosen for ecorestoration or social forestry, no adverse impacts on undergrowth or other nearby vegetation are evidenced.

Eucalyptus and Wild Life

There are several published studies on the behaviour of wild life in eucalyptus plantations, from Australia, Brazil, Malawi and South Africa. Two different kinds of comparisons are involved: between regions where eucalyptus is indigenous and the ones where it is an exotic, and between natural forests and plantations.

Views in Favour of Eucalypts

Comparison of natural (non-eucalypts) forest with plantations of eucalypts and of araucaria. Three papers cover this. Dietz *et al.* (1975) compared two areas of mixed natural forest with a 10-year old plantation of *E. saligna* and a 31 year old plantation of *Araucaria angustifolia,* a species native to Brazil. The two area had once been indigenous evergreen tropical rain forest and were recovering from complete devastation, one 15 years ago and the other 52. Populations of small mammals were sampled by trapping in all four forests. Five species of mammals were involved *(Orzymys nigripes, Monodelphis americana, Marmosa sp., Akodon arviculoides* and *Blarinomys breviceps).* The highest relative densities of small mammals were found in the araucaria plantation, the lowest in the eucalypts plantation. Densities in the two natural forests were statistically the same and lay between the two plantations. The diversity of captured species was highest in the natural forests and least in the homogeneous plantations.

Jocque (1981) compared the density of webs and the weight of the individuals of the large spider *Nephila sp.* in eleven plots each of 7 x 7 m in eucalypts plantation and in Brachystegia woodland. The density of webs, and average weight of spiders, was significantly higher in the latter, 950:200 webs per hectare and 958:770 mg weight. He concluded that this is due to on sufficient food in the plantation and surmises that the spider would disappear in large pure eucalypts plantations. The activity of termites was also significantly lower in the plantation. Steyn (1977) made a qualitative comparison of the bird populations between eucalypts plantations (mainly of *E. grandis*) which cover over 25,000 ha in north eastern Transvaal and the natural lowveld *(Lowveld Sour Bushveld)* containing strips of trees and shrubs along watercourse.

He studied the following aspects:

(a) those birds that use the eucalyptus plantation and which feed in the plantations;

(b) those that use the understory of the older plantations for breeding and hunting for food;

(c) those that breed in the plantations and feed elsewhere; and

(d) those that use eucalyptus plantations for feeding only.

Some lowveld birds also made irregular invasions into the plantations, particularly during a winter drought in 1969/70.

Steyn also identified some niches which were filled by specialised birds in Australia but which were not yet filled by indigenous species in South Africa.

He concluded that eucalyptus plantations were not as sterile and unsuitable for bird life was they were often accused of being. Some less adaptable species had been driven out; and plantations were unsuited to the way of life of some species—for example the purple crested louries *(Gallirex porphyreolophus),* which is a fruit-eater and widow birds, bishop birds and some larks and pipits which prefer open country.

Comparison on Indigenous Eucalyptus Forests with Eucalyptus Plantations

Woinarski (1979) compared the birds of a 25 year old plantation of *E. botryoides* with those of an adjacent natural forest of mixed age *E. dives* with an open, moderately tall canopy and a varied shrub and grass understory. The site was close to Melbourne, Australia. He found that six species were significantly more common in the plantation and nine species in the natural forest. The diversity of bird species was slightly higher in natural forest. There were more hawking species, fewer species that gleaned food from bark and branches, and fewer that are seeds, fruit and nectar. Some species were common in the interior of the plantations than at the edges.

Comparison of Indigenous Eucalyptus Forests with Forests of Pinus Radiata

Two comparison of this kind have been made. One, by Friend (1982) who worked at Gippsland, Victoria and compared mammals; the other by Neumann (1979) whose studies were based on beetles of North-east Victoria.

In case of mammals, the richness in species was lower and the proportion of introduced species was higher in eucalyptus plantations, particularly in the younger stands where no overstory vegetation existed. Certain small ground-dwelling species were favoured in such plantations (of particular ages), depending on their requirements for food and refuge. Larger ground-dwelling herbivores and carnivores were common in the pine forests, although feeding areas for herbivores were restricted in middle-aged trees to the edges of compartments or to tracks. Most arboreal herbivores, nectivores and users of tree hollows were uncommon in the plantations, where they were restricted to remnants of native forest. Their long-term survival in the plantations was considered questionable. Some aboreal species with relatively broad requirements for food and refuge were able to exist within older plantation compartments which supported some understorey shrubs.

In pine forests the number of mammal species was greatest near edges adjacent to native forest, and where there was a mosaic of remnants of native forest and pine stands of various ages.

The findings of Neumann (1979) on beetles were similar. The diversity of communities of beetles was found to be significantly higher in mature eucalyptus than in the older stands of pine because, in the eucalyptus, there was a more even distribution of individuals among species and a greater richness of species. In both, the range of species was greater during the spring and summer than during the autumn and winter.

Neginhal (1980) described the effects of progressive reafforestation since 1958 of secondary grassland in the Ranibennur Blackbuck Sanctuary of 119 km^2 in Karnataka, India. This resulted in the recovery of populations of the blackbuck *(Antilope cervicapra)*, Indian bustard *(Chloriotis nigriceps)* and wolf *(Canis lupus)* that were nearly extinct. He considered it doubtful whether this trend would be maintained if the remaining open areas were planted.

Karanth and Singh (1983) report that in Karnataka, India, population of black bucks have increased significantly since eucalypts plantation work started. There has been a marked increase in the population of Great Indian Bustard as well.

Kondas (1986) has reported that arboreal and ground birds are as abundant in eucalyptus plantations as in scrub lands.

Views Against

According to Foley and Bernard (1984), eucalypts is a threat to ecological stability. He is of the opinion that as they are often cultivated in large monocultural stand, the land in and around a eucalyptus plantation becomes almost devoid of local fauna.

Gupta (1986) has observed that eucalyptus flowers are not eaten by animals or birds as well as its thin canopy crown does not harbour birds or butterfly population. Hence he is of the opinion that the possibilities of these fauna's visit in eucalyptus plantation does not even arise at all.

According to Shiva and Bandyopadhyay (1987), eucalypts destroys the environment for soil fauna which are important "factories" for producing soil fertility and efficient "machines" for maintaining soil structure. They claim that scanty leaf litter of eucalypts is not effectively transformed into organic matter because eucalyptus is toxic to soil organisms constituting decomposer-food-chains. The earthworm *Lanipito mauriti* which is responsible for decomposition of leaf litter is found to be absent in eucalyptus plantations. Shiva and Bandyopadhyay also allege that eucalypts does not allow the presence of micro and macro avifauna because of its allelopathic effect.

On Balance

Whereas reports of systematic long-term experiments are available (as detailed in last but one section) to suggest that eucalypts are no more hostile to wildlife than other monocultures, similar reports of controlled experiments are not found to substantiate the case against eucalypts. Those who have written about unsuitability of eucalyptus to wild life have based their reports on visual observations made in heavily disturbed plantations. Further they seemed to have looked for the kind of biodiversity one finds in natural forests and, when not finding it in eucalyptus monocultures, have derided the latter without having checked how other monocultures perform in this respect.

The studies described in the last but one section bring forth the following observation.

(a) that plantations have a less diverse flora and fauna than indigenous forests;

(b) that plantations of exotics have a less diverse flora and fauna than plantations of indigenous species;

(c) that plantations can be made into more favourable habitats for animals and plants by appropriate management which provides suitable habitats for the species one wishes to attract; leaving patches or corridors of indigenous vegetation helps greatly in this;

(d) limited planting in treeless areas and the shelter that these provide can be beneficial to populations of wildlife.

In conclusion it may be pertinent to quote from Budowski (1984): *As to ecological deserts, it all depends on with what a eucalyptus plantation is compared. If it is with a nearby natural mixed forests, there is no doubt that the latter is much richer in fauna but if such a plantation is compared with a nearby scarcely covered slope as for instance a burnt over savanna, then it is highly probable that there is more animal life including nesting birds in the eucalyptus stand.*

In the present work several species of wild birds were spotted in the plantations of *Eucalyptus hybrid.*

Eucalyptus as Shelter-belts or Wind-breaks

Among the numerous uses to which eucalyptus has been put, is planting them to protect crops against strong winds.

Jenson (1983) report that they could not find any evidence to suggest that eucalypts differ in their effects as shelter-belts from any other trees.

Poore and Fries (1987) state that eucalypts are frequently planted as shelter belts and therefore provide some protection against wind erosion. However, they maintain, this is strictly a physical phenomenon, its effects depend solely upon the physical characteristics of the site and of the shelter belts. The species used has no influence on the result except of course different species have different physical characteristics.

There are no specific adverse reports against the use of eucalyptus as shelter belts except that, indirectly, they might harm the crops *via* their alleged allelopathic effects. The aspect of allelopathy has been discussed earlier and it can be safely concluded that all species of eucalypts, except the few which have recognized allelopathic effects, would make as good trees for shelter beds as any.

WHO IS LOBBYING FOR EUCALYPTUS?

Those who write or speak against eucalyptus often mention the 'eucalyptus lobby' and the 'conspiracy of the vested interests to hoist eucalyptus upon Indian lands'.

What is this lobby? What are its vested interests?

Eucalyptus is not a patented product; no single agency or group of agencies have ownership rights over eucalyptus.

Eucalyptus is different from such mass consumption item as pesticides, fertilizers, and medicines which one or other capitalist country could be allegedly manufacturing with the intension of marketing in India and amessing unholy fortune.

Then why someone would like to promote eucalyptus, less so thrust it, if one is aware of the so-called ecological dangers posed by eucalyptus?

After scanning through a vast body of anti-eucalyptus literature one is able to find only two arguments in which some specific reason is given for the predispositions towards eucalyptus; these arguments are:

(a) Foresters who are employed with the government support eucalypts because the genera is easier to tend and grow than other trees species proposed for social forestry. Eucalypts are far less favoured by cattle for browsing than other trees. This attribute as also the general hardiness of eucalypts makes it easier to raise them; in other words foresters get better results for their labours with eucalypts. There is also this hint that as foresters have to work less hard to establish eucalyptus; the tree provides 'an easy way out' for the allegedly lazy government employees.

(b) Industrialists need eucalyptus so they pressurise the government which in turn orders the forestry officials to promote eucalyptus.

In both the above-mentioned arguments, it appears, there are attempts to make flaws out of a virtue. That eucalyptus saplings are less prone to damage by cattle than other trees ought to be deemed as an *added* advantage and not the *sole* advantage.

As for the industrialist-government-forestry official nexus for promoting eucalypts it is a conjecture, though not a wholly unlikely one. But the point not to be overlooked is that as long as there are no unassailable grounds to shun eucalypts why the parties concerned should not promote it given its remunerativeness (which in turn is governed by market forces)? Even when there is no pressure from any government agency to plant eucalypts farmers go for them; even when other tree species are made available by social forestry initiatives, farmers prefer eucalypts.

3

A Profile of Neyveli Lignite Corporation and its Afforestation Programme

Neyveli Lignite Corporation (NLC), a public sector enterprise, is a large lignite-based thermal power plant situated near the semi-urban area of Neyveli in the South Arcot District of Tamil Nadu, India, between 11°33'-11°35'N and 79°28'-79°32'E at a distance of ~ 200 km South of Madras, India. The power generation at NLC is based on the exploitation and utilization of lignite, a low-grade fossil fuel.

NLC was constituted on 14th November, 1956, and started developing its first mine (Mine-I) in May, 1957. By now NLC has several mines and its campus is spread over an area of 480 km^2.

GEOLOGY, HYDROGEOLOGY AND TOP SOIL

Geology

The geological formation encountered in this area consists of miocene of the tertiary age, including argillacious and ferrogenous sand stones, clays, lignite and aquifer sand. The top portions of the sandstones are lateritic.

Hydrogeology

The lignite seam is underlain by artesian aquifer exerting pressure of 5 to 10 kg/cm^2. The static head is 15 to 35 m above Mean Sea Level (MSL). The Neyveli ground water basin is in the form of a syncline with a maximum thickness of about 400 m of water-bearing sand in the central portion dropping to 50 m in the West and East. The aquifer is about 50 km wide in the East-West direction and 60 km long between Gadilam and Vellam river. In order to mine lignite the aquifers have to be depressurized. They get recharged by precipitation falling over an area of 420 km^2 situated about 15 km North West of NLC. NLC being close to the Eastern coastal area and being in the cyclonic belt, a rainfall of about 650 to 1650 mm is recorded per annum. Huge volumes of storm water, recharge the water table. The ground water control is in operation since 1961 with pumping of about 55 mill/m^3 of water. Maximum pumpage of 120 MCM of water was achieved in 1964. The abstraction of water is necessary to depressurise the aquifer below the lignite.

Mine Overburden

The main overburden formations are ferrogenous sandstones and clays which form a porous solid like formation at the top of the lignite. The minerological constituents are predominantly quartz (38%) with a cementing medium (56%). The overburden thickness varies from 50 to 110 m. The external overburden dump has been raised by 40 to 54 m. The internal dump of overburden has been raised by about 15 to 50 m.

Climate

The data on various aspects of the climate of NLC campus (CARD, 1987) is presented below.

Temperature

Highest monthly average (maximum)	:	39°C
Lowest monthly average (minimum)	:	19.7°C
Highest yearly average (maximum)	:	33.7°C
Lowest yearly average (minimum)	:	23.3°C

Rainfall

Monthly (maximum)	:	750.9 mm
Monthly (minimum)	:	0 mm
Yearly (maximum)	:	1636.2 mm
Yearly (minimum)	:	676.2 mm

Humidity

Maximum	:	98%
Minimum	:	20%

Barometric Pressure

Maximum	:	764.80 mm of Hg
Minimum	:	747.02 mm of Hg

Wind Speed

Monthly (minimum)	:	Calm

NLC'S INDUSTRIAL ACTIVITIES

The main activity of NLC is power generation through lignite, a low-grade fossil fuel. Lignite is extracted from mines and is used in thermal power stations for generation of power. There are two thermal power stations *viz* TPS-I and TPS-II functioning as separate units in the same campus.

The lignite is also used in the Briquetting and Carbonisation Plant for the production of coke. The Fertilizer Plant uses low sulphur as the feed for the production of urea. The Clay Benefication Plant purifies raw white clay which is excavated from the mines during the extraction of lignite and is used in producing NEKOLIN which is used in porceilain, insulator, paper and rubber industries.

The products produced by various Plants/Mines and their annual capacity are as listed below:

Plant/Mine	Annual Capacity
Mine-I	6.5 mill tonnes of lignite/annum
TPS-I	600 MW of electricity/annum
Mine-II (Stage-I)	4.7 mill tonnes of lignite/annum
TPS-II (Stage-I)	620 MW of electricity/annum
Fertilizer Plant	1,29,200 tonnes of urea/annum
Briquetting and Carbonisation Plant	2,62,000 tonnes of coke/annum
Clay Washing Plant	6000 tonnes of washed clay/annum

As such the present thermal power capacity is 1220 MW/annum with a lignite mining at 11.2 tonnes/annum. With the turn of the century, NLC envisages increasing the capacity of power generation of 4200 MW/annum and lignite mining to 300 mill tonnes/annum.

Mining Operations

The type of mining practiced in NLC is opencast mining. After mining other activities which follows are drilling and blasting, and dumping.

Drilling and Blasting

Sixty per cent of the overburden require drilling and blasting.

Dumping

Dumping after mining is done on the open land covering an area of 195 hectares which is known as North-Dump. The bottom bench soil is cross dumped on the mine floor itself.

Equipment

The equipment commonly used for various mining activities include:

(i) Bucket Wheel Excavators (BWE) of 700 and 350 litres.

(ii) Mobile Transfer Conveyers (MTC) with 1000, 1200 to 1500 mm converyor belts.

(iii) Drills

(iv) Trippers

(v) Bench-lifts

(vi) Dozers and

(vii) Spreaders.

MINING AND ITS IMPACTS

Mining activities always have adverse effects on air, water, vegetation, *etc.* The mining area suffer mainly due to the digging of the soil, thus disturbing the physical, chemical, biological and aesthetic quality of the soil, which in turn lead to air pollution, water pollution, leaching of the soil and ultimately soil erosion. Unless these waste lands are not properly managed, the mining activities can play havoc with the surrounding environment. Hence, NLC have undertaken a major land reclamation programme.

Land Reclamation

To combat the adverse effects of mining and to promote environmental management programme, NLC has constituted an environmental management programme which can retain the land simultaneously with the mining operations. Their land reclamation programme include: Soil conservation, Development of parks and lakes, and Afforestation.

Soil Conservation

NLC has taken a number of measures to prevent soil erosion and wash-offs from overburden dumps. Garland drains around excavation sites, earth bunds around the top periphery of dumps, masonry chutes for lowering water from dump tops, check dams in gulleys formed on old dumps, and contour trenches in the dump slopes are the measures taken for soil conservation in NLC.

Development of Parks and Lakes

A number of parks have been developed over the reclaimed area and flowering trees have been planted in these parks. Ponds and lakes have also been created (Plate 1). Fish is cultured in these ponds and attempts are being made to provide an aesthetic look to these wastelands. The flowering trees have attracted a lot of birds and butterflies to these reclaimed lands.

Afforestation

Afforestation forms a major part of land reclamation programme in NLC. It aids in stabilising the soil, making the soil rich in nutrients and in re-establishing the pre-mining soil conditions (Plates 2-3). Table 1 presents the list of tree species planted and their total number in the afforestation programme. These afforested areas are mainly located at:

Plate 1. Park (background) and pond adding aesthetic look to the NLC campus.

Plate 2. Impact of mining.

(a) the top of outside dumping areas in Mine-I and II,

(b) slope of the dumps,

(c) at the foot of the dumps, and

(d) in addition, trees have also been planted on a large scale in the vacant and unused lands in the

Plate 3. A step towards ecorestoration.

township. The whole township sports a luscious green look with a soothing touch.

Table 1. Tree species employed in the afforestation programme

Sl. No.	*Botanical Name*	*Common Name*	*Total Number Planted*
1	2	3	4
1.	*Eucalyptus hybrid*	Mysore gum	16,28,420
2.	*Casuarina equisetifolia*	She oak	3,19,500
3.	*Dendrocalamus strictus*	Bamboo	1,69,400
4.	*Leucaena leucocephala*	Subabul	4,36,805
5.	*Anacardium occidentale*	Cashew	2,850
6.	*Acacia auriculiformis*	Pencil tree	5,16,445
7.	*Tamarindus indicus*	Tamarind	13,100
8.	*Azadirachta indica*	Neem	28,410
9.	*Acacia arabica*	Babul-black	92,700
10.	*Albizzia lebek*	Siris	43,940

(Table Contd.)

Table 1. Contd.

1	2	3	4
11.	*Tectona grandis*	Teak	11,160
12.	*Agave mexicana*	-	5,80,460
13.	*Samanea saman*	Rain tree	10,750
14.	*Delonix regia*	Gulmohar	4,920
15.	*Dalbergia sissoo*	Bombay rosewood	2,58,750
16.	*Terminalia arjuna*	-	70,450
17.	*Ceiba pentanbra*	White silk cotton tree	2,000
18.	*Melia dubia*	Himwood tree	700
19.	*Cedrela toona*	-	3,800
20.	*Grevillea robusta*	Silver oak	4,250
21.	*Parkinsonia aculaeta*	Jerusalem thorn	5,000
22.	*Peltophorum pterocarpum*	Copper pod	5,000
23.	*Acacia catechu*	Khair	13,000
24.	*Acacia leucopholoea*	Babul-white	56,500
25.	*Other fruit trees*	-	5,165
26.	*Pruspers spigeria*	-	10,000
27.	*Eucalyptus citriodora*	Blue-gum	7,000

NLC's Achievements

The tree population in NLC as on 31.3.88 stood at 11.15 millions (NLC Report, 1988 a). Major tree species include *Eucalyptus hybrid, Acacia auriculiformis, Leucaena leucocephala, etc* as enlisted in Table 1. The main reason for choosing these tree species were because of their speedy growth, soil binding property thereby stopping soil erosion and ability to grow in poor wasteland soil. The Horticulture Department of NLC has played a commendable role in the afforestation bid which won them the Indira Priyadarshini Vrikshamitra Award in the very first year of its inception in 1986 for maintaining ecological balance by adopting large scale tree planting. NLC has also won the FICCI award for the year 1987 for their commendable role in environmental preservation and pollution control and for having undertaken soil reclamation and afforestation programme. The water and air quality data (NLC Report, 1988 b) showed that the effluents are within the permissible limits, thereby upholding the credentials of NLC in the area of water pollution control and effluent treatment.

4

The Surveys

UNDERGROWTH SURVEY

The undergrowth survey was conducted during May-June of 1988 and during February-March of 1993, in the plots which occur in three typical zones.

(i) Mine-I

This zone is typified by very poor soil essentially consisting of mining overburden.

(ii) Township

This is the area characterised by good quality soil; well away from mining or industrial activity and free from frequent human intervention.

(iii) Thermal Power Stations I and II (TPS)

Areas in close proximity to TPS. This area is of good edaphic quality and lies between that of Mine-I and Township.

The surveyed plots comprised plantations of the following types:

(i) Mine-I

Eucalyptus hybrid (monoculture), *Leucaena leucocephala* (monoculture) and *Leucaena leucocephala—Anacardium occidentale—Prosopis juliflora* (mixed culture).

(ii) Township

Eucalyptus hybrid (monoculture), *Eucalyptus hybrid-Acacia arabica* (mixed culture), and *Eucalyptus hybrid—Acacia arabica—Prosopis juliflora* (mixed-culture).

(iii) TPS

Eucalyptus hybrid (monoculture) and *Acacia arabica* (monoculture).

The undergrowth found in the open land devoid of any tree plantation (control plot) in each of the three zones were also surveyed for comparison.

The survey (randomised quadrat-based enumeration) was conducted by systematic aligned sampling method (Williams, 1987). The quadrat sizes were set for each typical area on the basis of area-species diagrams. For obtaining these diagrams total number of species occurring in a randomly placed quadrat of 1 m^2 size were counted. Finally quadrat size and number of species were plotted on an x-y diagram (Figures 1-4) and optimal quadrat size was taken as the one beyond which increase in the quadrat size does not significantly increase the number of species encountered. In the present study the ideal quadrat size was found to be 3 m x 3 m. The area covered in each survey was

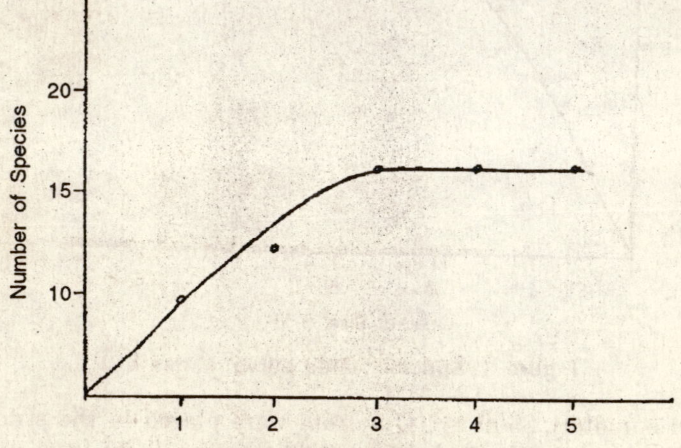

Figure 1: Species—area curve: TPI.

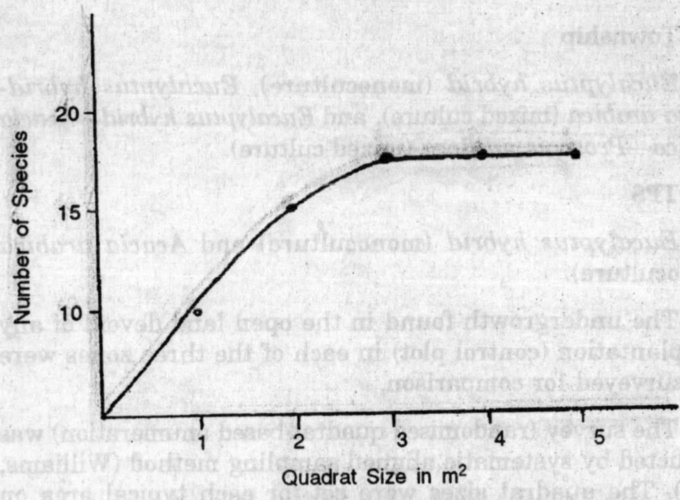

Figure 2: Species—area curve: Block V.

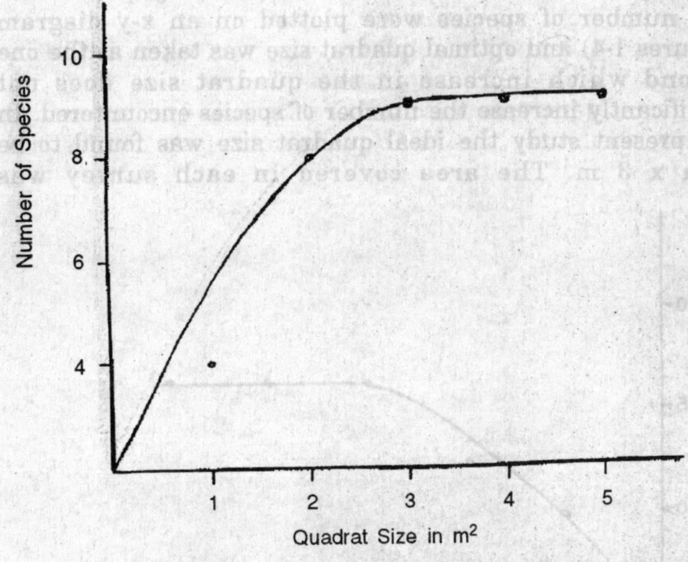

Figure 3: Species—area curve: Mines I.

approximately 2500 m^2. Quadrats were placed in the study plots in a hypothetical grid, equidistance (10 m) from one

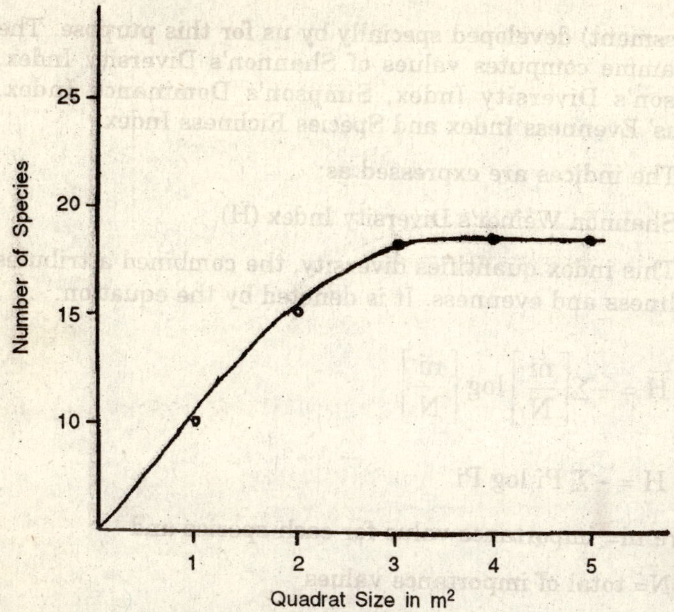

Figure 4: Species—area curve: Block VI.

another. In each quadrat the species encountered and the number of individuals of each species were noted. The method was an adaptation of the following procedure.

(a) Before sampling, the area to be studied has to be divided into sampling units of equal area.

(b) Decide the number of sampling points.

(c) Place these points regularly over the map.

(d) Each point on the actual sample area can be located either by placing or measuring with a tape. The quadrats are then kept by always locating the bottom left-hand corner of the quadrat at the point where the coordinates meet. The plants present at each quadrat are then identified and their numbers counted. Thus the species and their number encountered in the studies plots are recorded.

The data thus generated was analysed using a computer programme BIOMASS (BIOlogical Monitoring and

ASSessment) developed specially by us for this purpose. The programme computes values of Shannon's Diversity Index, Simpson's Diversity Index, Simpson's Dominance Index, Pielous' Evenness Index and Species Richness Index.

The indices are expressed as:

(a) Shannon Weiner's Diversity Index (H)

This index quantifies diversity, the combined attributes of richness and evenness. It is denoted by the equation:

$$\overline{\text{H}} = -\Sigma\left[\frac{\text{ni}}{\text{N}}\right]\log\left[\frac{\text{ni}}{\text{N}}\right]$$

$$\text{H} = -\Sigma \text{ Pi} \log \text{Pi}$$

Where ni= importance value for each species and

N= total of importance values

Pi= importance probability for each species

(b). Pielous's Evenness Index (e)

This index quantifies evenness in terms of the number of individuals per species.

$$(\text{e}) = \frac{\overline{\text{H}}}{\log \text{S}}$$

Where

$\overline{\text{H}}$ = Shannon Index

S = Number of species

(c) Simpson's Dominance Index (C)

This index quantifies the pre-dominance of a few species relative to other species. It is estimated with the formula:

$$\text{C} = \Sigma \text{ ni} \frac{\text{ni}(\text{ni}-1)}{\text{N}(\text{N}-1)}$$

(d) **Simpson's Diversity Index (cd)**

Is a diversity index denoted by the formula:

$$cd = 1 - \Sigma\,(ni/N)^2 \text{ and } \frac{1}{\Sigma(ni/N)^2}$$

(e) **Species Richness Index (d)**

This index quantifies species richness or the number of different species present. It is denoted by

$$d = \frac{S-1}{\log N}$$

Where S = No of species.

N = Total of importance values.

The dominance index indicates the predominance of a few species relative to other species. The richness index quantifies species richness or the number of different species present. The evenness index relates to the evenness in terms of the number of individuals per species. The diversity is a combined attribute of richness and evenness. Shannon's is one of the indices quantifying diversity.

An extensive study comparing the various indices was carried under six different plantation sites located in three different study areas as mentioned before.

SOIL CHARACTERISTICS

Sampling and analysis were done as per authentic soil analysis procedures (1988). From the plantations in each of the three zones namely Mine-I, Township, and TPS, soil samples were collected at three different depths (surface, ~ 7.5 cm, and 15 cm) and homogenized. In each zone, samples were also collected from a control area (area which is devoid of any tree plantation but otherwise similar to the plantation site in its soil status and other environmental conditions). The samples were preserved in 1 kg polythene bags. The soil in the bags was emptied on clean thick sheets of paper and evenly spread with a sampling knife. It was again heaped into

a cone by raising the four ends of the paper or using the knife, mixed well, and again spread evenly as before. The process was repeated three or four times to ensure uniformity and then finally spread evenly on the paper again. Now with the tip of the knife or a spatula, small quantities were taken at various points from the homogenized soil sample to get a representative sample for estimation.

The soil samples were analysed for estimating the moisture content, organic carbon content and nutrients (NPK) content in the soil in collaboration with CARD, NLC.

Estimation of Soil Moisture

The soil moisture content was estimated by the gravimetric method. Soil moisture was determined by heating a weighed quantity of soil (5 to 10 gm), in a well ventilated electric oven at 105°C to constant weight. The difference in the weight lost is taken as the direct measure of moisture content in the soil sample.

Estimation of Organic Carbon

Organic carbon was estimated by Walkely and Black method. The organic matter (humus) in the soil is oxidized by chromic acid (potassium dichromate plus concentrated sulphuric acid), utilising the heat of dilution of sulphuric acid. The unreacted dichromate is determined by back titration with ferrous ammonium sulphate.

Estimation of Nitrogen

Nitrogen was estimated by Alkaline Permanganate method. The method involves distilling the soil with alkaline $KMNO_4$ solution and determining the ammonia liberated which serves as an index of the available/minerabilisable nitrogen status.

Estimation of Phosphorous and Potassium

Phosphorous and potassium were estimated by citric acid extraction method. The method involves evoporating to dryness of two 500 ml portions of the filtrate separately and igniting the residue till all organic matter is destroyed. The ignited residue are digested with hydrochloric acid and evaporated to dryness. From this phosphoric acid was estimated as total P_2O_5 and Potash as K_2O by sodium cobalt nitrate method.

ORNITHOLOGICAL AND LEPIDOLOGICAL SURVEY

This particular study was carried out mainly at afforestation zone of the mining area because this was the area where most of the ecological impacts were felt due to mining. This study ascertains if the afforestation programme has encouraged the visits of birds and butterflies. These surveys were conducted by strip transect method (COX, 1990). Each study area was traversed in a *zig-zag* manner on several imaginary straight line paths (transects) running from end to end noting the species of the birds/butterflies encountered on the way. The *zig-zag* walk, at a speed of about 1 km/hr, terminated when no new species were sited even after traversing several new transects.

The method was an adoptation of the following procedure:

1. Select an area of habitat with conditions that are uniform for 125 m or more on either side of a transect route.
2. Mark the transect route at 100 m (or shorter) intervals for a distance of 1-2 km.
3. Familiarise yourself with the field marks, call notes and songs of one or more species that are well represented on the census area.
4. Conduct an actual census by walking slowly (1 km/hr in wood land, 2-3 km/hr in open country) along the transect route, pausing now and then to look and listen. In a note book, record each bird detected.

5

The Key Findings

UNDERGROWTH SURVEY

The survey led to the identification of 128 undergrowth species during 1988 and 198 species during 1993. There is, thus, a significant enhancement in the diversity of undergrowth over the years, indicating that the afforestation has had favourable ecological impact *vis a vis* undergrowth.

Eventhough we recorded 63.6% more species in 1993 compared to 1988, 17 species found during 1988 were not present during 1993 and at the same time there were 65 new species on 1993. This may be attributed to the fact that the 1988 survey was conducted during the summer (May-June) where as the 1993 survey was conducted soon after winter (February-March). Some of the seasonals that appear during mid-summer were, therefore, absent during the 1993 survey. Out of 181 species recorded 22 are leguminous of which 10 were reported in 1993 and other 19 in 1988. The species found during 1988 and the additional species found during 1993, their characteristics and potential uses, are listed in Tables 1 and 2 respectively. The list of species found during 1988 which were to encountered during 1993 survey is given in Table 3.

The undergrowth found during the two surveys in plots of different tree species, lying in there different areas of widely differing soil quality, is enumerated in Tables 4-6.

Among the total number of species recorded during 1988 and 1993 survey, the number of species exclusively found in

Table 1. Species of undergrowth found in 1988 in the afforested areas in Neyveli Lignite Corporation campus. Their characteristics and potential uses

Sl. No.	*Name of the Species*	*Frequency of flowering*	*Leguminous / Non-leguminous*	*Utility*	*Habit*
	1	**2**	**3**	**4**	**5**
1.	*Merremia angustifolia*	B	N.L	-	Herb (at twiner)
2.	*Borreria hispida*	P	"	-	Herb
3.	*Syzygium cumini*	-	"	Wood used for building purposes	Tree
4.	*Pavonia zeylanica*@	A	"	-	Herb
5.	*Solanum surattense**	-	"	-	"
6.	*Alternanthera pungens*	A	"	Garden borders as its leaves turn a crimson colour	"
7.	*Sida cordifolia*	"	"	-	-
8.	*Bridelia retusa**	"	"	Bark, grey to olive to brown with pretty silver grain useful and durable wood for furniture *etc.*	Tree

	1	2	3	4	5
9.	*Clitorea ternata*	-	L	Cultivated in gardens, selfsown in hedges and thickets	Shrub (Tendril) climber)
10.	*Triumfetta rhomboidea*	A	N.L	-	Herb
11.	*Moringa oleifera*	-	"	Roots have flavour of horse radish. Seeds eaten in curries and give a valuable oil	Tree
12.	*Psidium guajava*	-	"	Fruits edible and very tasty. Bark and leaves are used for tanning and dyeing. Wood is made into spear-handles and special instruments	"
13.	*Lannea coromandelica*	-	"	Young leaves eaten with rice. Wood pulp used for preparing paper and board. Gum obtained by tapping stem bark	"

				is used in confectionery, calicoprinting paper and cloth sizing and as mucilage for making ink. Wood is also used for furniture, agricultural implements hair brushes packing cases, tea chests, slates frames, carving, turnery and also suitable for commercial plywood writing and printing paper	
14.	*Asystasis gangetica*	A/B	"	Ornamental with pure blooms. Grows well in rockeries	Herb
15.	*Andrographis echioides*	A	"	Drug from an inflor-escence used as tonic for fevers, worms, dysentery and liver and digestive complaints	"
16.	*Abrus precatorius*	-	L	Seeds hard and used for jewellers weights for necklaces and other ornaments.	Shrub and Climber

	1	2	3	4	5
17	*Alangium salvifolium*	P	N.L	Scented. Useful for ornamental work and is a good fuel	Tree
18.	*Argyreia pilsoa*	A	"	Ornamental (Climber)	Shrub
19.	*Blastania garcinic*	"	"	Orange-red fruits edible	Herb
20.	*Enterlobium saman*	"	L	Fruits used as fodder	Tree
21.	*Euphorbia hirta**	"	N.L	Entire plant drug for bronchial infections, cough, asthma and removing intestinal worms from children. Latex of plant applied in warts	Herb
22.	*Alternanthera sessils*	"	"	Ornamental: found in garden borders	"
23.	*Xanthium strumarium*	"	"	Used as organic manure used as medicine for chronic malaria and urinary troubles. Seed-oil is edible and also employed in various industries. Fruits rich is vitamin C	"

	1	2	3	4	5
24.	*Cyperus spp*[*]	A/P	"	Seen in rice fields and marshy places. Valuable pasture grass	Sedge
25.	*Casuarina equisetifolia*	-	"	Planted as avenue tree used for fuel. Used for reclaiming sand dunes. Wood is used for transmission poles and for paper making. Bank is used as tan	Tree
26.	*Cassyta filiformis*	-	"	Parasitic plant. Used in bilious afflictions, chronic dysentery and skin diseases	Twig
27.	*Amaranthus spinosus*[#]	A	"	Leaves sometimes eaten as spinach	Herb
28.	*Emblica officinalis*	-	"	Fruits edible. Used for tanning. Bark is used for poles, implements and furniture	Tree

	1	2	3	4	5
29.	*Flacourtia ramontchi*	-	"	Fruits edible. Used for making jams, jellies, syrups and preservatives	Shrub
30.	*Hyptis suaveolens**	A	"	Aromatic and used as medicine for common cold, cough and fever	Herb
31.	*Languiera spp*	-	-	-	-
32.	*Lepidagathis cristata*	P	N.L	Plant is used as bitter tonic is fevers. Also applied to itchy infections of skin. Leaves are used as fodder	Herb
33.	*Lippia nodiflora*$	"	"	An alcoholic extract of leaves show antibacterial activity against *E. coli* and contain 8% tannin. Plant yields 2 glucoside colouring matter Nodiflorin -A and Nodiflorin-B	Herb/ Shrub
34.	*Ocimum sanctum*	"	"	Leaf juice is medicinal Used for fever cough and cold. The oil glands have medicinal values. It is a sacred plant for Hindus and is seen in temples	Herb

	1	2	3	4	5
35.	*Oldanlandia umbellata*	B	”	Liliac flowers are of considerable value, bark and root yield valuable red dye	”
36.	*Pongamia pinnata*	-	L	Wood used for cart wheels. Seed give and oil used for burning and in medicines	Tree
37.	*Peltophorum pterocarpum*	-	L	Ornamental with bright yellow flowers	Tree
38.	*Parthenium hysterophorus**	-	N.L	Usedastonic, febrifuge and emmenagogue; decoction of root is given in dysentery	Herb
39.	*Phyllanthus madaraspatensis*	P/A	”	An infusion of leaves is used for headache. Seed possess laxative, carminative and diuretic properties	Herb
40.	*Sebastiana chamaelea**	A	”	Ornamental in cultivated lands	”
41.	*Tragia involuctra*	P	”	-	”

	1	2	3	4	5
42.	*Waltheria indica*@@	A	”	Medicinal use	under shrub
43.	*Zizypus jujuba*	-	-	Fruits edible. Used for salads, and is a rich source of vitamin C and sugars. Bark is used for tanning and leaves are used as fodder. It is a good fuel	Shrub or spiny tree small
44.	*Zornia diphylla*	A	L	Used as fodder for cattle	Wiry herb
45.	*Vernonia dalgelliana*	-	N.L	Fencing work	”
46.	*Boerhaevia diffusa**	-	”	Give pwaarravine drug, also used for asthma	”
47.	*Justicia simplex*	-	”	-	”
48.	*Abutilon indicum*	-	”	Ornamental, stem used for setting	”
49.	*Sida acuta*	-	”	Stem fibre used for ropes, leaves for ileumatic infections	”

	1	2	3	4	5
50.	*Lantana crenulata*	-	"	Used for fencing	Shrub
51.	*Fluggea microcarpa*	-	"	-	"
52.	*Tephrosia purpurea*	-	L	User as green manue, leaves yield dye	Herb
53.	*Mimosa pudica*	-	"	Used as a green manure and is ornamental Leaves yield dye	"
54.	*Calotropis procera*	More than once a year	N.L	Stem and leaves as green manure, root and bark for dysentry	Shrub
55.	*Gomphrena decumbens*	-	"	-	Herb
56.	*Indigofera tinctoria*	-	"	Juice of leaves used for hydrophobia, epilepsy and in bronchitis	Shrub
57.	*Crotalaria juncea*	P	L	Stemfiber for nets, mats cigarettes paper. Leaves as manure	"
58.	*Jatropha curcas*	-	N.L	Seed oil has purgative property	"
59.	*Phaseolus mungo*	-	L	Fodder crop	"

	1	2	3	4	5
60.	*Opuntia dillenii*	P	N.L	As hedge plant	"
61.	*Leucas aspera*	A	"	-	"
62.	*Hybanthas enneaspermum*	-	"	-	Herb
63.	*Cassia siamea*	P	"	Ornamental flower	Sapling
64.	*Croton bonpliandanum*		"	-	Herb
65.	*Anotis wightiana*	P	"	-	Villous
66.	*Cissampelos pareira*	-	"	-	Tomentose climber
67.	*Lantana indica**	-	"	Planted in gardens	Shrub
68.	*Cassia occidentalis*	P	N.L	-	Herb
69.	*Echinochloa colona*	P/A	"	Excellent fodder grain eaten by poorer classes	-
70.	*Cassia alata*	-	L	Ornamental and also used for treating skin diseases	-
71.	*Vernonia fysoni*	P	N.L	-	Shrub
72.	*Tridax procumbens**	-	"	-	Herb

	1	2	3	4	5
73.	*Clausena wildinovii*	-	”	Fruit edible and worth cultivation	Small tree
74.	*Achyranthes aspera*	-	”	-	Herb
75.	*Evolvulus alsinoides*	P	”	-	Herb
76.	*Cissus quadrangularia**	-	N.L	Heading factured bones	Succu- -lent Twiner
77.	*Hemidesmeus indicus*	-	”	-	Twining Shrub
78.	*Chloris barbata*	P	”	Fodder	Grass
79.	*Cynodon dactylon*	-	”	Fodder and commonest	”
80.	*Chloroxylon swietenia*	-	”	For making paper	Tree sapling
81.	*Solanum xanthocarpum*	-	”	-	Herb
82.	*Calotropis gigantica**	-	”	Bark gives fiber and silky material for pillows	Milky shrub
83.	*Doclonaea viscosa*	-	”	Firewood, flowers leaves give tannin used as hedge plant	Shrub

	1	2	3	4	5
84.	*Stylosanthes fruticosa*	-	"	For asthma, diarrhea and dysentery	Climber
85.	*Arundo donax*	P	"	Cullms used for baskets, Papers and rayon grade pulp	-
86.	*Prosopis juliflora**	"	L	Firewood for fencing etc.	Grass
87.	*Achrocephalus indicus*	-	_	-	Shrub
88.	*Ipomea biloba*	-	N.L	Good land binder	Runner
89.	*Phyllanthus niruri**	-	"	Native medicine for	Herb
90.	*Dendrocalamus strictus*	-	"	Culms used for poles	Grass
91.	*Gmelina asiatica*	-	"	Planted as hedge	Shrub
92.	*Vitex trifolia*	-	"	-	Herb
93.	*Cassia tora*	P	L	Seeds used for preparation	"
94.	*Scoparia dulcis*	-	N.L		"
95.	*Cassyta filiformis*	-	"	Used in bilious affliction	Twiner
96.	*Pergularia extensa*	P	"	-	"
97.	*Merremia tridentata*	"	"	Used in rheumatism, piles and skin disease	Herb
98.	*Polygala chinensis*	-	"	Leaves for asthma	
99.	*Mollugo pentaphylla*	P	N.L	Medicinal	Herb
100.	*Mollugo oppositifolia**	"	"	-	"

	1	2	3	4	5
101.	*Aniseia ceniflora*	-	"	-	Prostrate Herb
102.	*Asristida adscencionis*	A	"	-	Tufted Herb
103.	*Coldenia procumbens*	-	N.L	Stems lie flat on the ground	Herb
104.	*Delonix regia*	-	L	Grown as avenue trees	Tree
105.	*Tylophora indica*	P	"	Leaves are used to treat asthma	Climber
106.	*Eragrostis spp*	A	N.L	-	Herb
107.	*Todalia asiatica*	-	"	The root bark yields yellow dye	Shrub
108.	*Typha latifolia*	P	"	-	Herb
109.	*Canthium angustifoliam*	-	"	-	-
110.	*Canthium parviflorum*	-	"	Stems and branches are largely used as fencing materials	-
111.	*Phoenix spp*	-	"	Sweet pulp of seeds eatable, leaflets used for mats and thatching and petioles for basket making	Shrub
112.	*Morinda tinctoria*	-	"	-	Decid-uous tree

	1	2	3	4	5
113.	*Fimbristylis meliaceae*	-	"	-	Herb
114.	*Fimbristylis schoechoides*	-	"	-	"
115.	*Securiniga leucopyrus*	-	"	-	Rigid bush
116.	*Stachytarophetta jamaicensis*	-	"	-	Herb
117.	*Azadirachta indica*	-	"	Leaves use as disinfectant	Tree
118.	*Citrullus colocynthis*	-	"	Fruit pulp affords important purgative medicine	Trailing Herb
119.	*Dalbergia sissoo*	-	L	Woods used for making furniture, improves sterile lands, can withstand drought	Tree
120.	*Acacia auriculiformis*	-	"	-	Tree
121.	*Albizia odorotissima*	-	"	Heart wood very hard and is largely used for cartwheels	Tree

	1	2	3	4	5
122.	*Cassia auriculata*	-	"	Bark is used for tanning and dyeing leather. The seeds, barks and leaves as medicinal	Shrub
123.	*Eragrotis pilosa*	A	N.L	-	Herb
124.	*Randia spp*	-	"	Leaves used for mats thatching houses. Fed to buffaloes	-
125.	*Justicia montana*	P	"	-	Shrub
126.	*Saccharum spp*	"	"	Leaves are used for mats and thatching houses, and fed to buffaloes	Herb
127.	*Leucaena leucocephala*	-	L	Wood used as fuel. Leaves use as fodder and manure	Tree
128.	*Ludwigia parviflora**	-	N.L	-	Herb

B = Biennial
P = Perennial
A = Annual
N.L = Non-leguminous
L = Leguminous
* = Weedy
@ = Uncommon weed
# = Troublesome weed
$ = Weedy in wet grounds and pastures
@@ = Weedy (common).

Table 2. Species of undergrowth found only in 1993 in the afforested areas in Neyveli Lignite Corporation. Their characteristics and potential uses

SL. No.	*Name of the Species*	*Frequency of flowering*	*Leguminous / Non-leguminous*	*Utility*	*Habit*
	1	**2**	**3**	**4**	**5**
1.	*Cardiospermum halicacabum*	-	N.L	-	Herb (a twiner)
2.	*Mollugo nudicalis*	A	N.L	-	Herb
3.	*Desmodium laburnifolium*	-	L	-	Shrub
4.	*Carrisa spinarum*	-	N.L	-	Thorny shrub
5.	*Justicia glauca*	-	N.L	-	Herb
6.	*Mimosa dichrostachis*	-	L	-	-
7.	*Crotalaria retusa*	-	L	Herbaceous shrub giving fibres	Shrub
8.	*Fluggea leucopyrus*	-	N.L	-	-
9.	*Solanum torvum*	-	N.L	Fruits are edible	Shrub
10.	*Physalis minima**	A	N.L	-	-
11.	*Ipomea dissecta*	-	N.L	-	-

	1	2	3	4	5
12.	*Memecylon talbatianum*	P	N.L	-	Tree
13.	*Barleria prionitis*	-	N.L	-	Shrub
14.	*Acalypha indica**	A	N.L	Medicinal	Herb
15.	*Melothria madraspatana*	-	N.L	-	Climber
16.	*Clausena heptaphyalla*	-	N.L	Aromatic	Tree
17.	*Passiflora foetida*	-	N.L	-	Herb
18.	*Cassia fistula*	P	L	Bark is smooth when young, and tough when old. Hard wood, strong and durable, especially used for agricultural work	Tree
19.	*Vernonia anamalica*	-	N.L	-	Shrub
20.	*Crotalaria laevigata*	-	L	-	Shrub
21.	*Polygala javana*	-	N.L	-	Shrub
22.	*Jasminum malabaricum*	-	N.L	-	Shrub (Climber)

	1	2	3	4	5
23.	*Cadaba indica*	-	N.L	-	Shrub
24.	*Desmodium biarticulum*	-	L	-	Shrub
25.	*Cayratia auriculata*	-	N.L	-	Climber
26.	*Eragrotis cynosuroides*	-	N.L	-	-
27.	*Eleucine indica*	-	N.L	-	Herb
28.	*Ipomea tridentata*	P	N.L	-	Herb
29.	*Blepharis umbellata*	-	N.L	-	-
30.	*Alysicarpus monilifer*	P	L	-	Herb
31.	*Daemia extensa*	-	N.L	-	Climber
32.	*Glycosmis cochinchinensis*	-	N.L	-	Shrub
33.	*Stylosanthes hemata*	-	L	-	-
34.	*Kirgenelia sp*	-	N.L	-	Shrub
35.	*Pavett indiga*	-	N.L	-	-
36.	*Polycarpon cosimboson*	-	N.L	-	Herb
37.	*Vernonia setigera*	-	N.L	-	Shrub

	1	2	3	4	5
38.	*Carrisa carandas*	-	N.L	Berries are edible, wood is used to make combs and spoons and is used as fuel. The branches are used for fencing	Shrub (Thorny)
39.	*Hedyotis sp*	-	N.L	-	Herbs
40.	*Corchorus accutangulus*	-	N.L	-	-
41.	*Lonicera ligustrina*	-	N.L	Used as hedge plant	Shrub
42.	*Albizia amara*	P	L	Strong wood, used in agricultural and building purposes	Tree
43.	*Desmodium triflorum*	-	L	-	Herb
44.	*Hibiscus ovilifolium*	A/B	N.L	-	Herb
45.	*Sporobolus indicus*	-	N.L	-	Grass
46.	*Polygala rosmarinifolia*	A	N.L	-	Herb
47.	*Zizyphus oenoplia*	-	N.L	Fruits are edible and the branches are used for fencing	Shrub
48.	*Passiflora edulis*	-	N.L	-	Shrub

	1	2	3	4	5
49.	*Lycospersicon esculentum*	-	N.L	Commonly cultivated for fruit and occasionally run wild	Shrubby herbs
50.	*Indigofera cordifolia*	A	L	-	Prostrate
51.	*Ionodium suffruticosum*	-	N.L	-	Under shrub
52.	*Evolvulus numellaris*	P	N.L	-	Herb
53.	*Alysicarpus monilifer*	P	L	-	Herb
54.	*Ehretia miocrophylla*	-	-	-	Shrub
55.	*Marsilia spp*	-	-	-	-
56.	*Acacia arabica*	-	L	Pods eaten by cattle, wood for agricultural and many other purposes	Moderate sized tree
57.	*Ficus bengalensis*	-	N.L	Planted in avenues and shades. Wood used for well curbs, tent and yokeploes	Large tree
58.	*Ocimum cannum*	-	N.L	Flowers used in medicine	Erect herb

	1	2	3	4	5
59.	*Wrightia tinctoria*	-	N.L	Leaves give a blue dye and wood used for carving	Deciduous tree
60.	*Melothria purpusilla*	-	N.L	-	Climber
61.	*Morinda angustifolia*	-	N.L	-	Small tree
62.	*Ocimum basilicum*	-	N.L	Aromatic plant used for culinary seasoning of fish, meats and salads	Herb
63.	*Commelina persicariaefolia*	-	N.L	-	Herbs
64.	*Aristida setaceae*	P	N.L	Used for making brooms	-
65.	*Ulteria salicifolia*	-	-	-	-
66.	*Moschosma polystachyum*	A	N.L	-	Herb
67.	*Imaridium indicum*	-	-	-	-
68.	*Spermaceae hispida*	-	-	-	-

NL = Non-leguminous
L = Leguminous
P = Perennial
B = Biennial
A = Annual

Table 3. Species of undergrowth found in 1988 which were not encountered during 1993 survey

Sl. No.	*Name of the species*	*Habits*	*Uses*
1.	*Cloture ternatea*	of Table 1,	Species number 9
2.	*Asystasia gangetica*	of Table 1,	Species number 14
3.	*Alangium salvifolium*	of Table1,	Species number 17
4.	*Blastania garcini*	of Table 1,	Species number 19
5.	*Xanthium strumarium*	of Table 1,	Species number 23
6.	*Cassyta filiformis*	of Table 1,	Species number 26
7.	*Lippia nodiflora*	of Table 1,	Species number 33
8.	*Waltheria indica*	of Table 1,	Species number 42
9.	*Jatropha glandulifera*	of Table 1,	Species number 58
10.	*Phaseolus mungo*	of Table 1,	Species number 59
11.	*Cissampelos pareira*	of Table 1,	Species number 66
12.	*Lantana indica*	of Table 1,	Species number 67
13.	*Cassia occidentalis*	of Table 1,	Species number 68
14.	*Echinochloa colona*	of Table 1,	Species number 69
15.	*Achrocephalus indicus*	of Table 1,	Species number 87
16.	*Vitex trifolia*	of Table 1,	Species number 92
17.	*Aniseia ceniflora*	of Table 1,	Species number 101

Eucalyptus hybrid plantations were 22 and 55 during 1988 and 1993 respectively (Table 7). The number of species exclusively found in mixed plantations with *Eucalyptus hybrid* as one of the tree species were 7 and 53 during 1988 and 1993 respectively (Table 8). The number of species exclusively found in monocultures other than *Eucalyptus hybrid* were 3 during 1988 as well as in 1993 (Table 9). The number of species exclusively found in mixed plantation of trees other than *Eucalyptus hybrid* were nil during 1988, and during 1993 one species was recorded (Table 10). The number of species found common to all plantations were 2 and 7 during 1988 and 1993 respectively (Table 11). Six species were found

Table 4. Undergrowth in afforested areas of Mine-I

Plot	*Code in Map-1*	*Year of plantation*	*Planted species*	*Number of species of undergrowth*		*% increase in the number of undergrowth species*
				1988	*1993*	
Plot 13	A	1983	*E. hybrid*	17	18	5.88
Plot 5	B	1982-83	*E. hybrid*	9	16	77.75
Plot 16	D	1982-83	*E. hybrid*	7	12	71.42
Plot 11A	F	1982-83	*E. hybrid*	8	12	50.00
Plot 8	G	1982-83	*E. hybrid*	6	16	166.00
Plot 15	H	1983	*E. hybrid*	6	16	166.00
Plot 1	I	1982-83	*L. leucocephala*	5	12	140.00
Plot 11B	J	1983-84	*Mixed (E. hybrid and A. auriculiformis)*	4	9	125.00
Plot 6	K	1982	*L. leucocephala, T. grandis, A. occidentale and C. equisetifolia*	10	23	130.00
Control	C1	-	-	-	10	-

Table 5. Undergrowth in afforested areas of Township

Plot	*Code in Map-1*	*Year of plantation*	*Planted species*	*Number of species of undergrowth*		*% increase in the number of undergrowth species*
				1988	*1993*	
Block 5	L	1985	*A. auriculiformis, E. hybrid and P. juliflora*	36	47	30.55
Near first mine road	N	1983	*E. hybrid*	71	73	2.81
Plot 13	R	1983-84	*E. hybrid*	17	43	152.94
Plot 14	C	1982-83	*E. hybrid*	18	32	77.77
Block 6a	T	1985-86	*E. hybrid and A. auriculiformis*	23	42	82.60
Block 6	U	1984-85	-do-	16	30	87.50
Block 24	V	1985	*E.hybrid, A. auriculiformis and P. juliflora*	34	36	5.88
Control	C3	-	-	-	33	-

Table 6. Undergrowth in afforested areas of TPS region

Plot	*Code in Map-1*	*Year of plantation*	*Planted species*	*Number of species of undergrowth*		*% increase in the number of undergrowth species*
				1988	*1993*	
Opposite TPS - I	M	1983-84	*E. hybrid*	52	48	-7.70
Plot 4	P	1984-85	*E. hybrid*	30	39	30.00
Near Pipe line	O	1985-86	*A. auriculiformis*	28	30	7.14
Near helipad	Q	1987	*E. hybrid*	23	31	34.80
Plot 2	S	1934-85	*E. hybrid*	23	43	87.95
Control	C2	-	-	-	20	-

Table 7. Species exclusively found in *Eucalyptus hybrid* plantations during 1988 and 1993 survey

Sl. No.	*1988*	*1993*
1	**2**	**3**
1.	*Ocimum sanctum*	*Cardiospermum helicacabum*
2.	*Parthenium hysterophones*	*Physalis minima*
3.	*Sebastiana chamaelia*	*Ipomea dissecta*
4.	*Tragia involucrata*	*Memecylon talbatianum*
5.	*Cynodon dactylon*	*Acalypha indica*
6.	*Solanum xanthocarpum*	*Melothria madraspatana*
7.	*Calotropis gigantia*	*Passiflora foetida*
8.	*Prosopis juliflora*	*Cassia fistula*
9.	*Cassyta filiformis*	*Vernonia anamalica*
10.	*Merrimia tridendata*	*Cadaba indica*
11.	*Polygala chinensis*	*Eragrotis cynosuroides*
12.	*Agave*	*Elecucine indica*
13.	*Syzegium cumini*	*Stylosanthes haemata*
14.	*Sida cordifolia*	*Kirginelia sp.*
15.	*Psidium gujava*	*Pavetta indica*
16.	*Blastinia garcini*	*Corchorus accutangulus*
17.	*Enterlobium saman*	*Lonicera linguistrina*
18.	*Lantana camara*	*Lycopersicon esculentum*
19.	*Acacia auriculiformis*	*Ionodium suffroticosm*
20.	*Leucaena leucocephala*	*Evolvulus numellaris*
21.	*Acalypha indica*	*Morinda angustifola*
22.	*Cassia tora*	*Ocimum basilicum*
23.	—	*Aristida setaceae*
24.	—	*Spermaceae hispida*
25.	—	*Solanum surattense*
26.	—	*Alternanthera pungens*
27.	—	*Bridelia retusa*
28.	—	*Lannea coromandalica*
29.	—	*Argeyra pilosa*
30.	—	*Alternanthera sessils*
31.	—	*Hyptis suaviolens*
32.	—	*Parthenium hysterophones*
33.	—	*Sebastiana chemelea*

1	2	3
34.	—	*Zornia diphylla*
35.	—	*Hybanthes enreaspermum*
36.	—	*Cassia siamia*
37.	—	*Vernonia fysoni*
38.	—	*Cissus quadrangularia*
39.	—	*Cynodon dactylon*
40.	—	*Solanum xanthocarpum*
41.	—	*Gmelia asiatica*
42.	—	*Scoparia dulcis*
43.	—	*Delonix regia*
44.	—	*Morinda tinctorea*
45.	—	*Securiniga leucopyrus*
46.	—	*Citrulus colosinthes*
47.	—	*Dalbergia sissoo*
48.	—	*Leucaena leucopholia*
49.	—	*Ludwigia parviflora*
50.	—	*Mimosa pudica*
51.	—	*Jatropha gossipifolia*
52.	—	*Micrococceae mercurialis*
53.	—	*Datura metal*
54.	—	*Pavonia zeylanica*
55.	—	*Scoparia dulcis*

Table 8. Species exclusively found in mixed plantations of *Eucalyptus hybrid* and other tree species during 1988 and 1993 survey

Sl. No.	*1988*	*1993*
1	2	3
1.	*Phaseolus mungo*	*Mallugo nudicalis*
2.	*Opuntia dillenii*	*Carrisa spinarum*
3.	*Cissampelos pareira*	*Solanum torvum*
4.	*Stylosanthes fruticosa*	*Dalbergia prionitis*
5.	*Chomelia asiatica*	*Clausena heptaphylla*
6.	*Desmodium laburnifolium*	*Polygala javana*
7.	*Desmòdium biarticulatum*	*Jasminum malabaricum*
8.	—	*Desmodium biariculum*
9.	—	*Castya auriculata*

1	2	3
10.	—	*Hybiscus oviliformis*
11.	—	*Polygala rosmarinifolia*
12.	—	*Passiflora edulis*
13.	—	*Ehritia microphylla*
14.	—	*Marsila sp.*
15.	—	*Acacia arabica*
16.	—	*Ocimum cannum*
17.	—	*Wrightia tinctoria*
18.	—	*Melothria purpusilla*
19.	—	*Commelena persicariafolia*
20.	—	*Ulteria salicifolia*
21.	—	*Moschosma polystachyum*
22.	—	*Imaradium indicum*
23.	—	*Syzigium cumini*
24.	—	*Andrographis echioides*
25.	—	*Abrus precetorious*
26.	—	*Cyperus sp.*
27.	—	*Casurarina equisetifolia*
28.	—	*Amaranthus spinosus*
29.	—	*Emblica officinalis*
30.	—	*Languria sp.*
31.	—	*Lepidogathis cristata*
32.	—	*Peltophorum pterocarpum*
33.	—	*Phyllanthus madraspatensis*
34.	—	*Tragia involucrata*
35.	—	*Zyziphus jujuba*
36.	—	*Vernonia dalgelliana*
37.	—	*Justicea simplex*
38.	—	*Lantana crenulata*
39.	—	*Crotolaria junceae*
40.	—	*Anotis wightiana*
41.	—	*Cassia alata*
42.	—	*Calotropis gigantia*
43.	—	*Ipomea biloba*
44.	—	*Phyllanthus niruri*
45.	—	*Dendrocalamus strictus*

1	2	3
46.	—	Cassyta filiform
47.	—	Pergularia extena
48.	—	Polygala chinensis
49.	—	Mollugo oppositifolia
50.	—	Coldenia procumbens
51.	—	Tylophora indica
52.	—	Justicea montana
53.	—	Phaseolus mungo

Table 9. Species exclusively found in monoculture plantations other than *Eucalyptus hybrid* during 1988 and 1993 survey

Sl. No.	*1988*	*1993*
1.	*Cissus quadrangularis*	*Moringa olerifera*
2.	*Chloris barbata*	*Cassia occidentalis*
3.	*Toddalia asiatica*	*Agave mexicana*

Table 10. Species exclusively found in mixed plantations of other tree species during 1988 and 1993 survey

Sl. No.	*1988*	*1993*
1	*Nil*	*Flacourtia romantchi*

exclusively in *Eucalyptus hybrid* plantations both during 1988 and 1993 where as, one species was exclusively found in other monoculture plantations both during 1988 and 1993 (Table 12).

Table 11. Species found common to all plantations during 1988 and 1993 survey

Sl. No.	*1988*	*1993*
1.	*Tephrutia purpurea*	*Merremia angustifolia*
2.	*Evolvulus alsinoides*	*Sida acuta*
3.	—	*Tephrutia purpurea*
4.	—	*Dodonia viscosa*
5.	—	*Mollugo pentaphylla*
6.	—	*Fimbristyles schonoides*
7.	—	*Vernonia cineria*

Table 12. Species exclusively found in *Eucalyptus hybrid* plantations both during 1988 and 1993 survey

Sl. No.	*Name of the Species*
1.	*Parthenium hysterophones*
2.	*Sebastiana chamaelia*
3.	*Cynodon dactylon*
4.	*Solanum xanthocarpum*
5.	*Leucaena leucoecephala*
6.	*Acalypha indica*

A large number of species found by us have recognised uses as medidicinals, ornamental plants, sources of aromatics, fodder, and edibles (Gamble, 1935), as given in Tables 1 and 2. Some of the undergrowth species encountered during the survey are shown in plate 1. These species help in holding the soil particles together, thus reducing a chance of soil erosion. Joshie and Narain (1994), Mathur *et al* (1980) and Bhasa (1986) have also reported a reduction in run-off and soil loss, due to excellent undergrowth allowed by the typically open canopy of *Eucalyptus hybrid* plantation. The

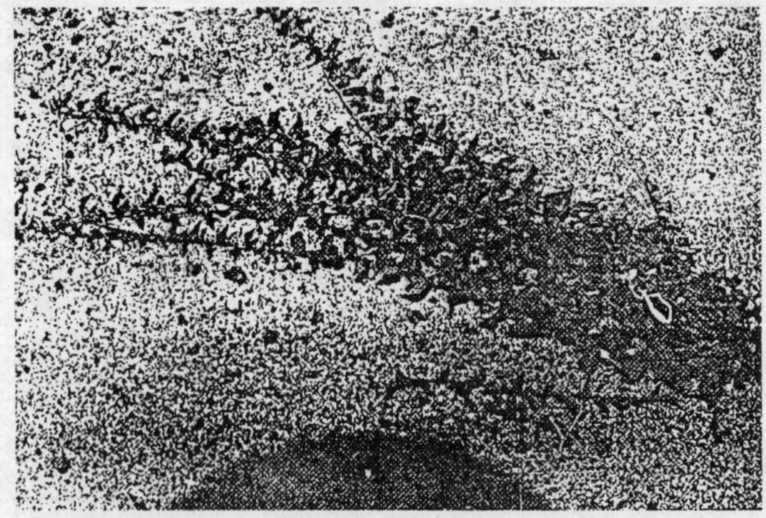

Vernonia tridendata

Cassia tora

Plate 1. Two of the undergrowth species occurring naturally under *Eucalytus hybrid* plantation.

leguminous plants help in nitrogen fixation. The plants attract insects and birds in large numbers, and signs of pollination are visible as indicated by the presence of other tree saplings in the monoculture plots. The number of undergrowth species is more in the township than in the mines area. This could be because the lands afforested in the township area were only wastelands where as the mines afforestation programme was carried on the overburdens when the land was in a disturbed state.

Based on the above findings it can be assumed that effects of *Eucalyptus hybrid* on the undergrowth were in no way harmful. The performance was always comparable with the of other tree plantations. The presence of undergrowth in *Eucalyptus hybrid* plantations has also been reported earlier by Bhasa (1986), Gupta (1986) and Kushalappa (1985).

Undergrowth Diversity Indices

To put the findings on a firm quantitative footing and to compare the results, a very extensive study comparing the various indices *viz* Simpson's diversity index, Shanon Weiner's diversity index, Simpson's dominance index, Species richness index and Pielous's evenness index was carried out in the six different plantation sites located in three different study areas as mentionex in chapter 4.

The index values of the studied plots are presented in Table 13.

(i) Simpson's diversity index

(a) **Township.** During 1988, *Eucalyptus hybrid* plantations has greater diversity than other plantations, but during 1993, *Eucalyptus hybrid-Acacia auriculiformis* mixed plantation is ahead of the rest followed by *Eucalyptus hybrid* monoculture plantation (Figure 1).

(b) **TPS**. During 1988, the performance of *Eucalyptus hybrid* monoculture plantation easily matches with that of *Acacia auriculiformis* monoculture plantation but in 1993 *Eucalyptus hybrid* monoculture plantation outperforms the rest (Figure 2).

(c) **Mine-I**. The performance of *Eucalyptus hybrid* monoculture plantation is always comparable with that of other mono and mixed cultures (Figure 3).

(ii) Shannon Weiner's diversity index

(a) **Township.** *Eucalyptus hybrid* monoculture plantation has the highest diversity of undergrowth (Figure 4).

(b) **TPS.** During 1988, *Eucalyptus hybrid* monoculture plantation evenly matches with that of *Acacia auriculiformis* monoculture plantation. During 1993, the undergrowth diversity in *Eucalyptus hybrid* monoculture plantation is next to *Acacia auriculiformis* monoculture plantation (Figure 5).

(c) **Mine-I.** *Eucalyptus hybrid* monoculture plantation has distinct edge over the rest (Figure 6).

Table 13. Index values of undergrowth vegetation in the sample plots

Plantation	*Location*	*Species Richness*	*Shanon Weiner's Diversity*	*Pielous's Eveness*	*Simpsons's Dominance*	*Simpsons's Diversity*
1	**2**	**3**	**4**	**5**	**6**	**7**
E. hybrid	Mine-I	3.6636	2.6636	0.0946	0.080	0.0644
E-hybrid	Mine-I	2.9430	2.3673	0.9527	0.1032	0.0813
E. hybrid and A. anriculiformis	Mine-I	2.8264	2.3601	0.9498	0.1054	0.0867
Mixed	Mine-I	2.3297	2.1084	0.9596	0.1322	0.1032
E. hybrid	Mine-I	3.3502	2.5360	0.9147	0.0997	0.0893
E. hybrid	Mine-I	3.1850	2.5651	0.9252	0.0921	0.839
E. hybrid	Mine-I	4.1867	2.6577	0.9195	0.0852	0.0672
L. leucocephala	Mine-I	2.6161	2.4172	0.9728	0.0805	0.0942
L. leucocephala, A. auriculiformis and C. equisetifolia	Mine-I	4.9390	2.9761	0.9492	0.0582	0.0473
Control	Mine-I	1.6987	2.0897	0.9075	0.1409	0.1366

Table 13. Contd.

1	2	3	4	5	6	7
A. auriculiformis	TPS-II	6.0264	3.0575	0.8990	0.0617	0.0540
E. hybrid	TPS-II	5.4405	2.6535	0.7243	0.0021	0.1029
E. hybrid	TPS-II	5.2560	2.4293	0.7142	0.1433	0.1398
E. hybrid	TPS-II	7.0701	3.3628	0.8687	0.0517	0.0505
E. hybrid	TPS-I	5.9025	2.9355	0.7805	0.0860	0.0853
Control	TPS-II	3.0217	2.642	0.8820	0.0852	0.0835
E. hybrid	Township	6.5208	2.8426	0.7558	0.1296	0.1282
E. hybrid	Township	4.7369	2.8260	0.8229	0.0972	0.0956
E. hybrid, A. auriculiformis and P. juliflora	Township	5.1273	3.0842	0.8746	0.0634	0.0619
E. hybrid	Township	10.6516	3.7207	0.8677	0.0447	0.0430
E. hybrid and A. auriculiformis	Township	6.1858	2.080	0.7708	0.0969	0.0946
E. hybrid, and A. auriculiformis	Township	6.0990	3.1011	0.8465	0.0655	0.0617
E. hybrid, A. auriculiformis and P. juliflora	Township	7.0545	3.0640	0.7958	0.0716	0.0703
Control	Township	4.7934	3.0774	0.8801	0.0640	0.0626

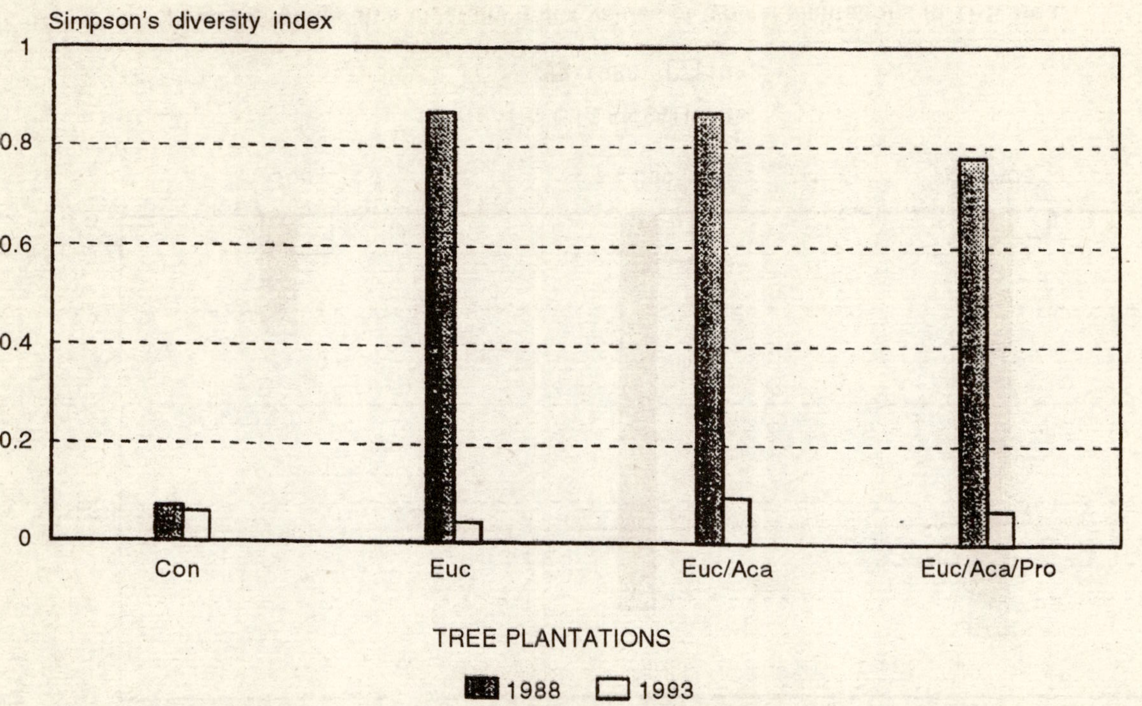

Figure 1. Simpson's diversity index values of typical plantations in township area.

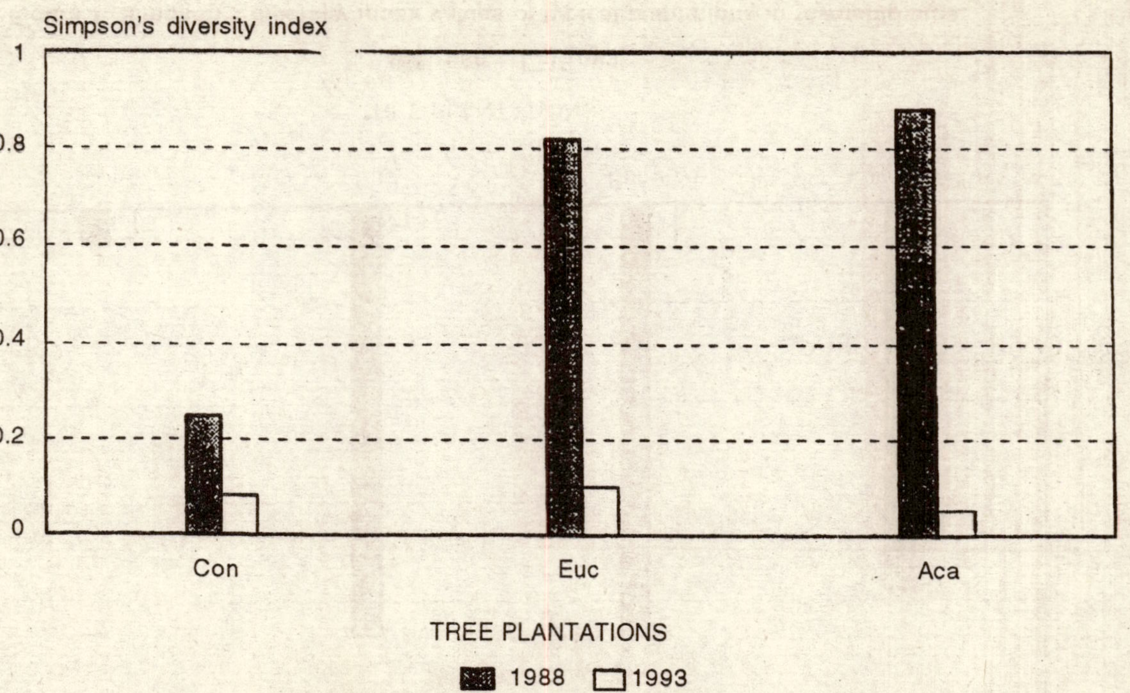

Figure 2. Simpson's diversity index values of typical plantations in TPS area.

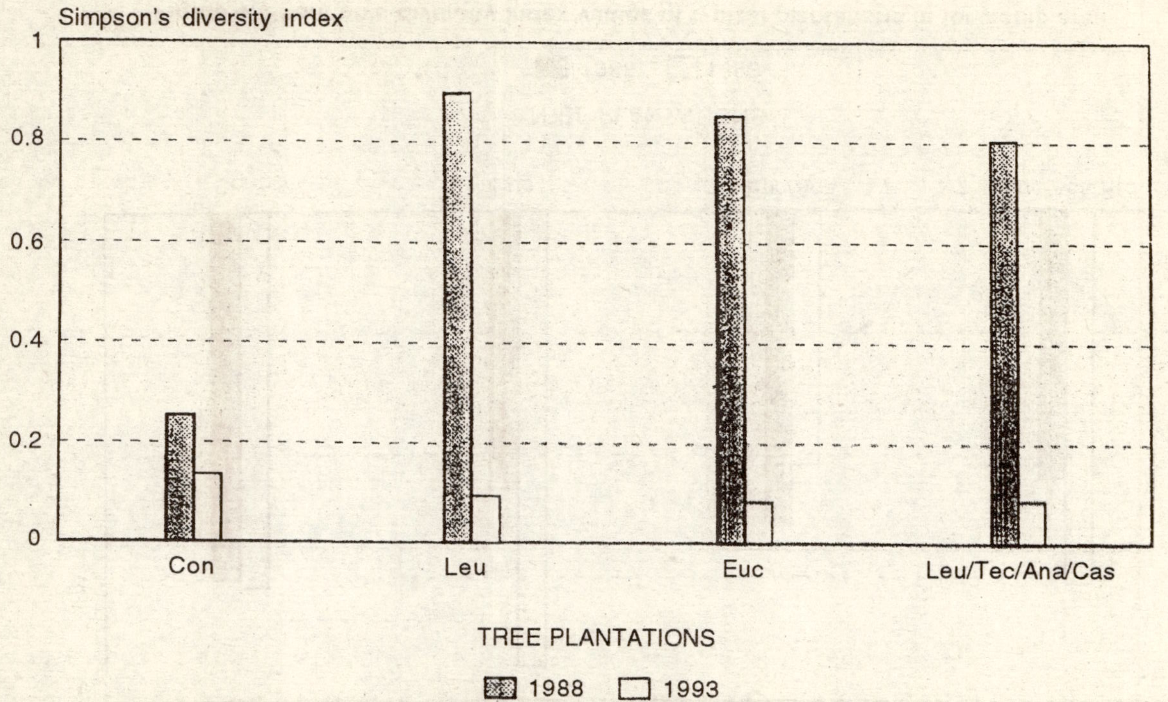

Figure 3. Simpson's diversity index values of typical plantations in mining area.

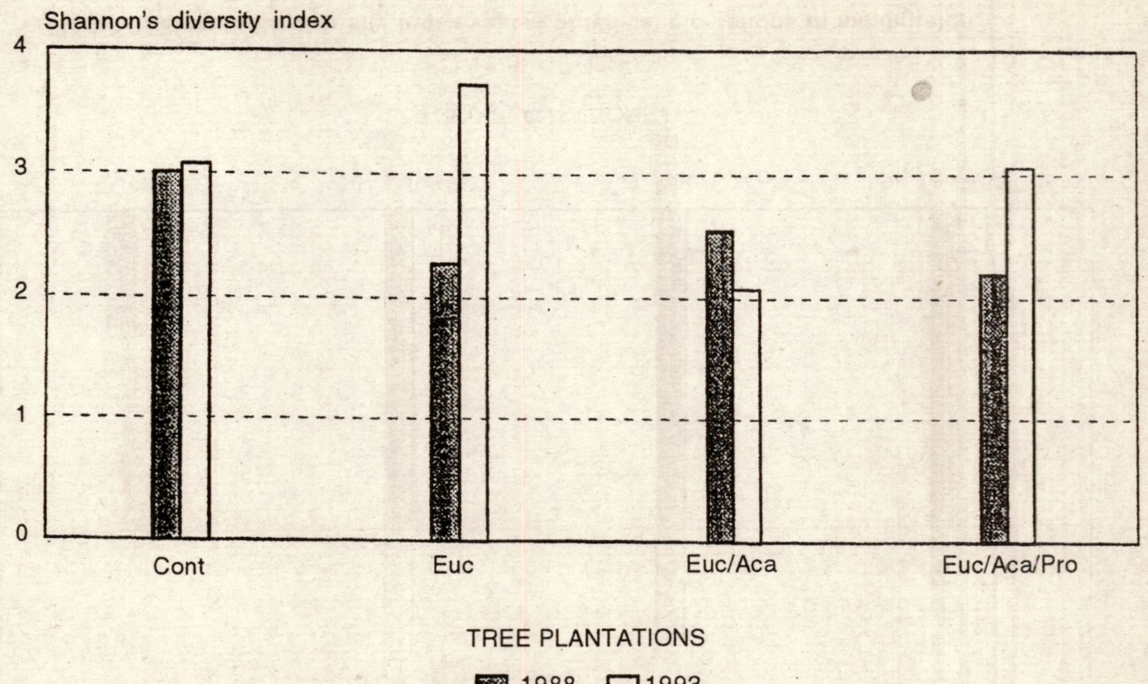

Figure 4. Shannon's diversity index values of typical plantations in township area.

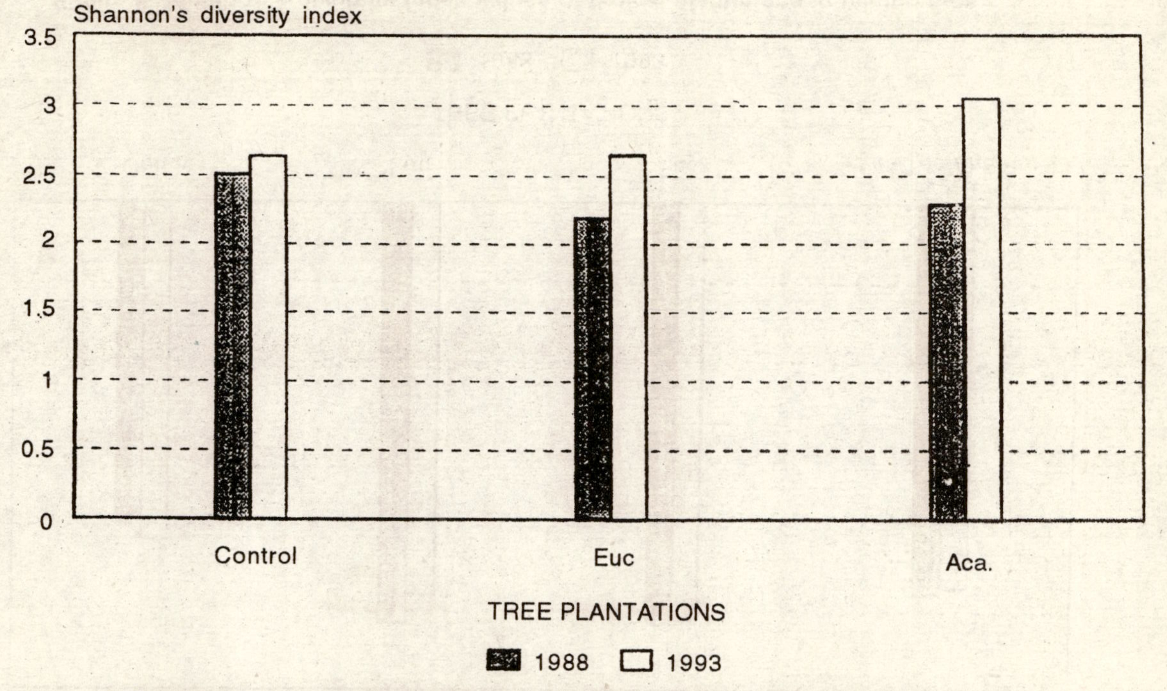

Figure 5. Shannon's diversity index values of typical plantations in TPS area.

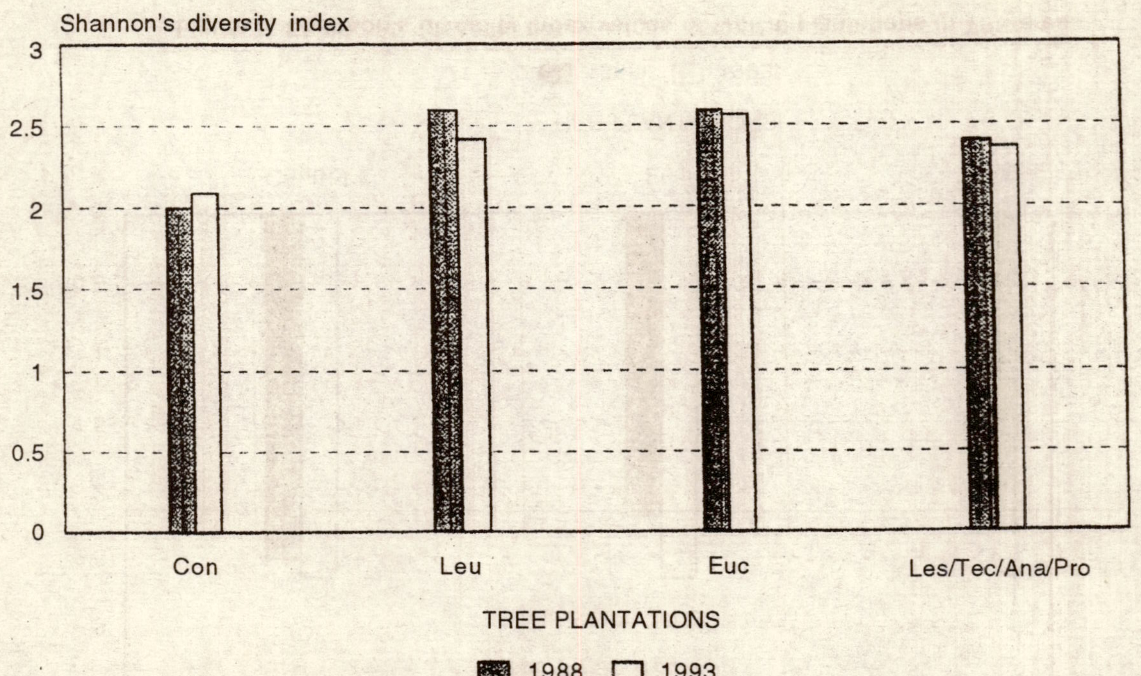

Figure 6. Shannon's diversity index values of typical plantations in mining area.

(iii) Simpson's dominance index

(a) Township. During 1988, undergrowth dominance was greater in *Eucalyptus hybrid—Acacia auriculiformis—Prosopis juliflora* mixed culture plantation followed by *Eucalyptus hybrid* monoculture plantation. During 1993, a sharp decline in undergrowth dominance has occurred in all the plots irrespective of the type of plantation (Figure 7).

(b) TPS. During 1988, the dominance of the undergrowth in *Eucalyptus hybrid* monoculture plantation was greater than in the *Acacia auriculiforms* monoculture plantation. During 1993 a sharp decline in undergrowth dominance has occurred in both the plantations (Figure 8).

(c) Mine-I. During 1988, undergrowth dominance was greater in *Eucalyptus hybrid* monoculture plantation but during 1993 undergrowth dominance is greater in *Leucaena leucocephala—Anacardium occidentale—Prosopis juliflora* mixed culture plantation (Figure 9).

(iv) Pielous eveness index

(a) Township. Though during 1988 the eveness of the undergrowth in *Eucalyptus hybrid* monoculture plantation was next to *Eucalyptus hybrid-Acacia auriculiformis* mixed culture plantation, during 1993, it has improved over the rest (Figure 10).

(b) TPS. The evenness of the undergrowth is more in the *Acacia auriculiformis* monoculture plantation than in the rest of the plantation (Figure 11).

(c) Mine-I. Both during 1988 and 1993, the evenness of the undergrowth in *Eucalyptus hybrid* monoculture plantation is comparable with that of other mono and mixed cultures (Figure 12).

(v) Species richness index

(a) Township. During 1988, *Eucalyptus hybrid—Acacia auriculiformis* mixed culture plantation is ahead of

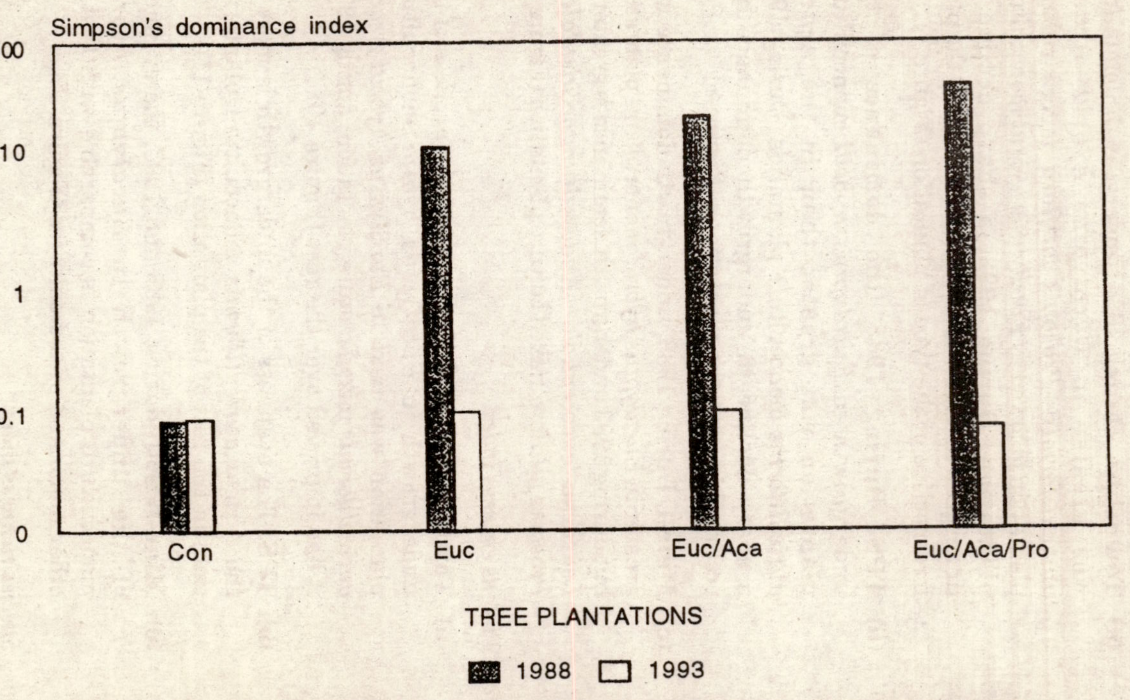

Figure 7. Simpson's dominance index values of typical plantations in township area.

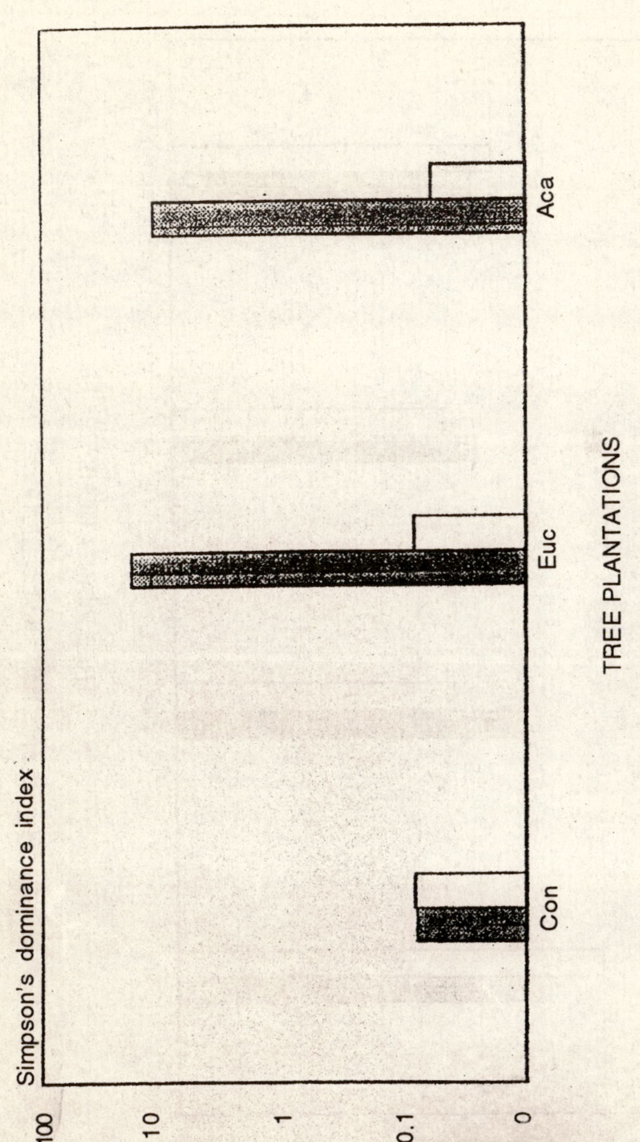

Figure 8. Simpson's dominance index values of typical plantations in TPS area.

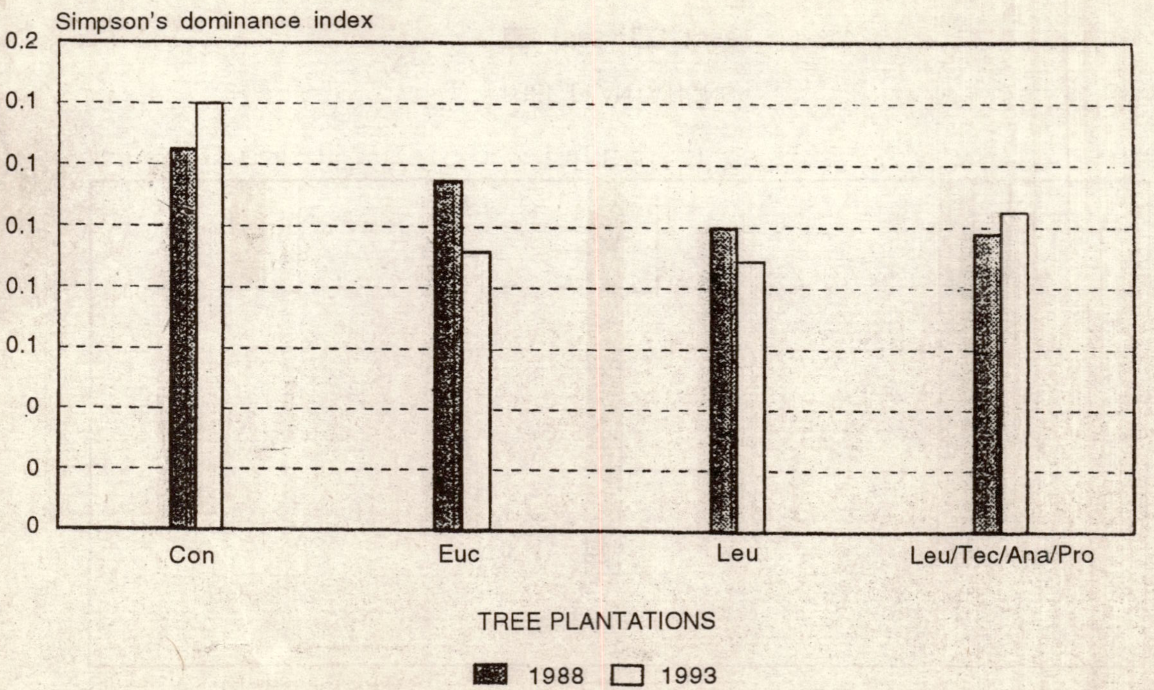

Figure 9. Simpson's dominance index values of typical plantations in mining area.

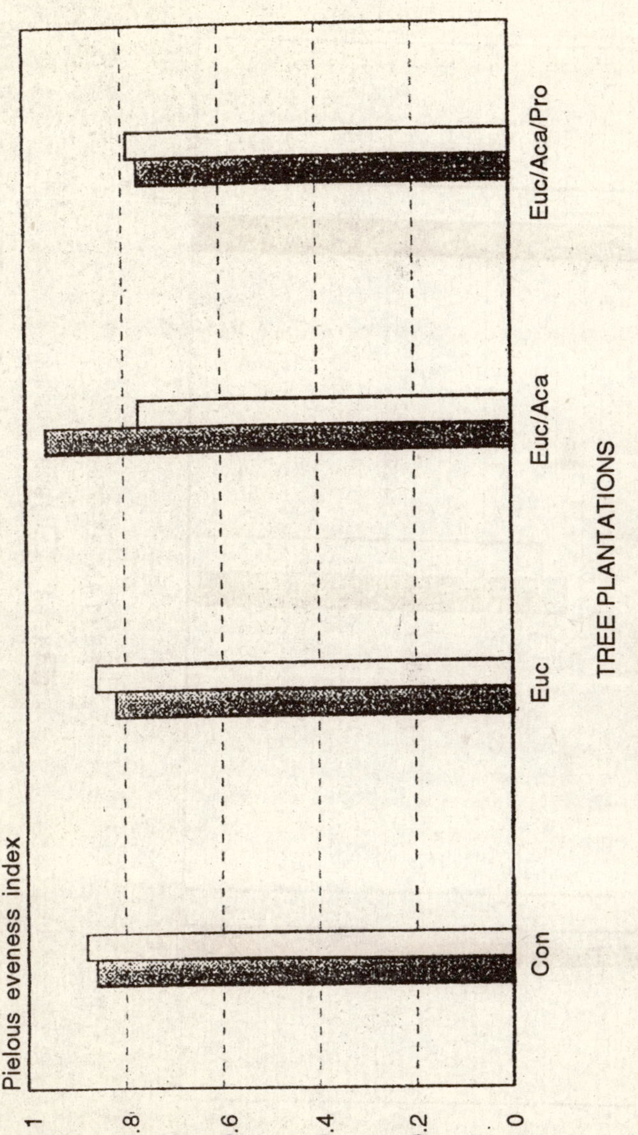

Figure 10. Pielous's eveneness index values of typical plantations in township area.

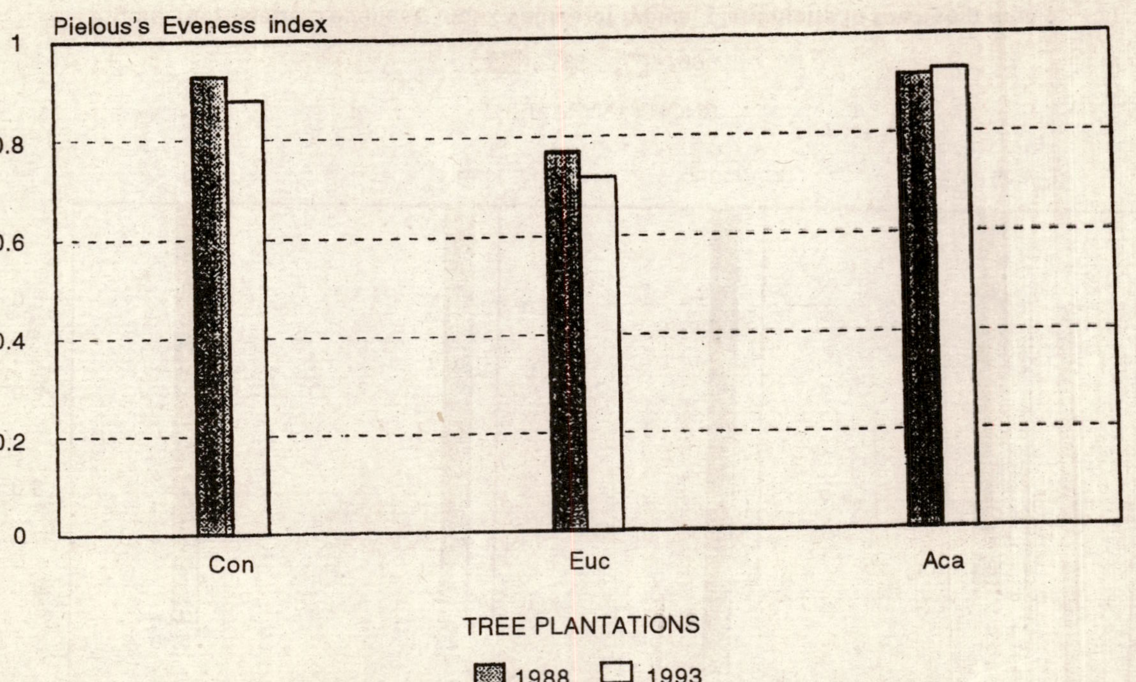

Figure 11. Pielous's eveness index values of typical plantations in TPS area.

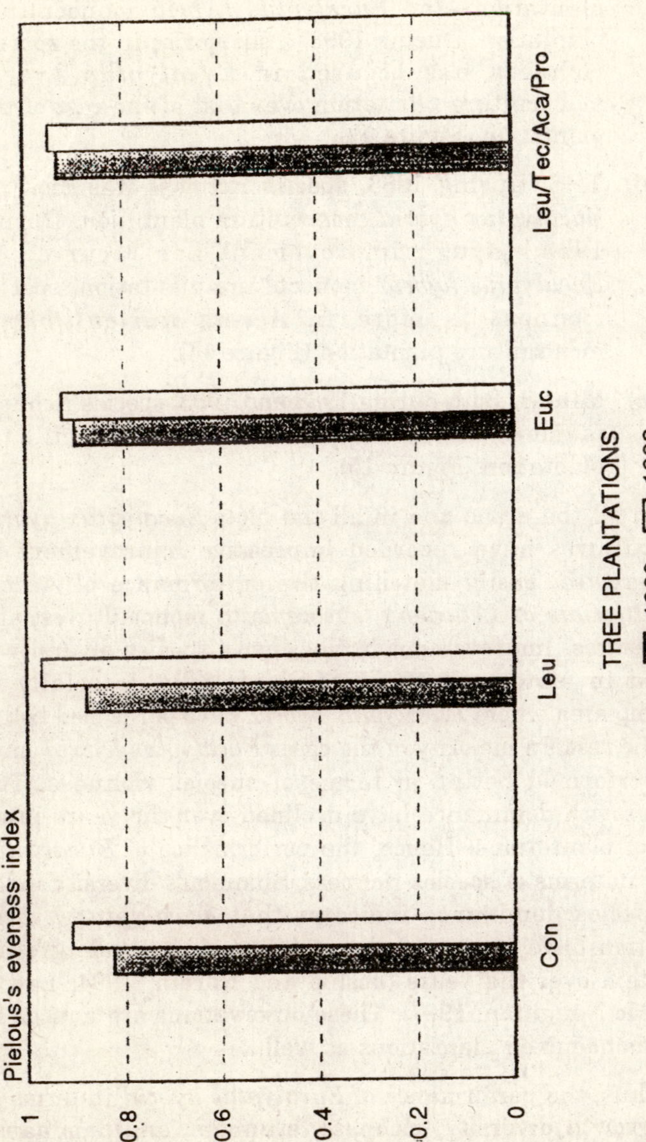

Figure 12. Pielous's eveneness index values of typical plantations in mining area.

the rest followed by *Eucalyptus hybrid—Acacia auriculiformis—Prosopis juliflora* mixed culture plantation, and *Eucalyptus hybrid* monoculture plantation. During 1993, a sharp rise in the species richness has occured in *Eucalyptus hybrid* monoculture plantation over that of the rest of the plantation (Figure 13).

(b) **TPS.** During 1988, species richness was more in *Eucalyptus hybrid* monoculture plantation. During 1993, though improvement has occurred in *Eucalyptus hybrid* monoculture plantation, species richness is more in *Acacia auriculiformis* monoculture plantation (Figure 14).

(c) **Mine-I.** Both during 1988 and 1993, species richness is more in *Eucalyptus hybrid* monoculture plantation (Figure 15).

In all the areas and in all the plots *Eucalyptus hybrid* monocultures have recorded impressive improvement in undergrowth, easily matching the performance of *Acacia auriculiformis* or *Leucaena leucocephala* monocultures, and polycultures. Improvement in Shannon's species diversity is noticed in almost all types of plantations especially in township area where *Eucalyptus hybrid* have performed better than the rest. In majority of the cases *Eucalyptus hybrid* have also performed better in terms of species richness. The undergrowth dominance have declined over the years in all types of plantations. Hence, the performance of *Eucalyptus hybrid* in terms of species richness, Shannon's diversity index and species dominance indicates that *Eucalyptus hybrid* plantation have increased the stability of the undergrowth ecosystem over the years (Joshie and Narain, 1994; Lewis, 1970; Mc Naughton, 1967). These observations are noticed in other mono/mixed plantations as well.

Thus, the performance of *Eucalyptus hybrid* in terms of undergrowth diversity, richness, evenness, and dominance have either showed impressive improvement or were always comparable with the performance of other tree plantations.

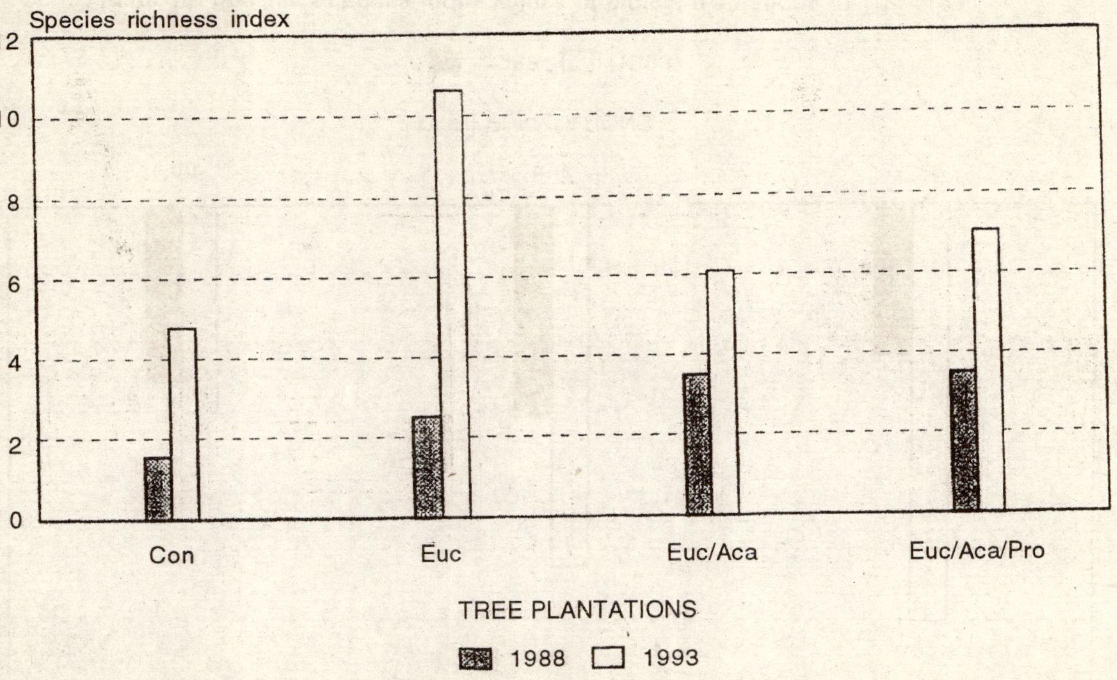

Figure 13. Species richness index values of typical plantations in township area.

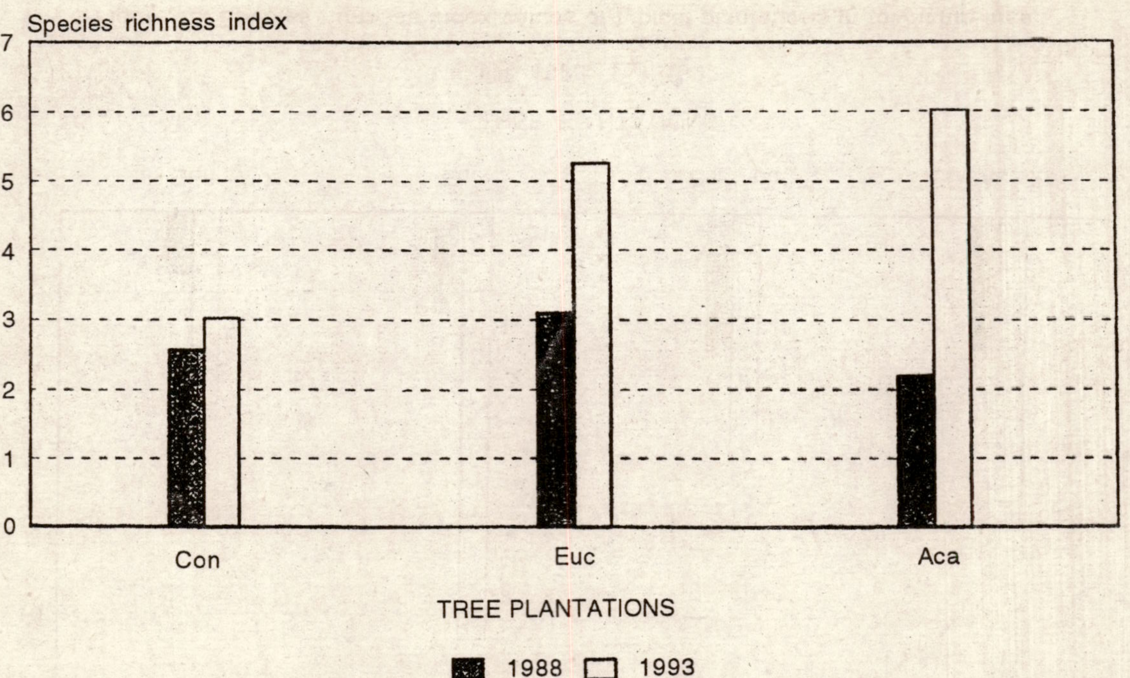

Figure 14. Species richness index values of typical plantations in TPS area.

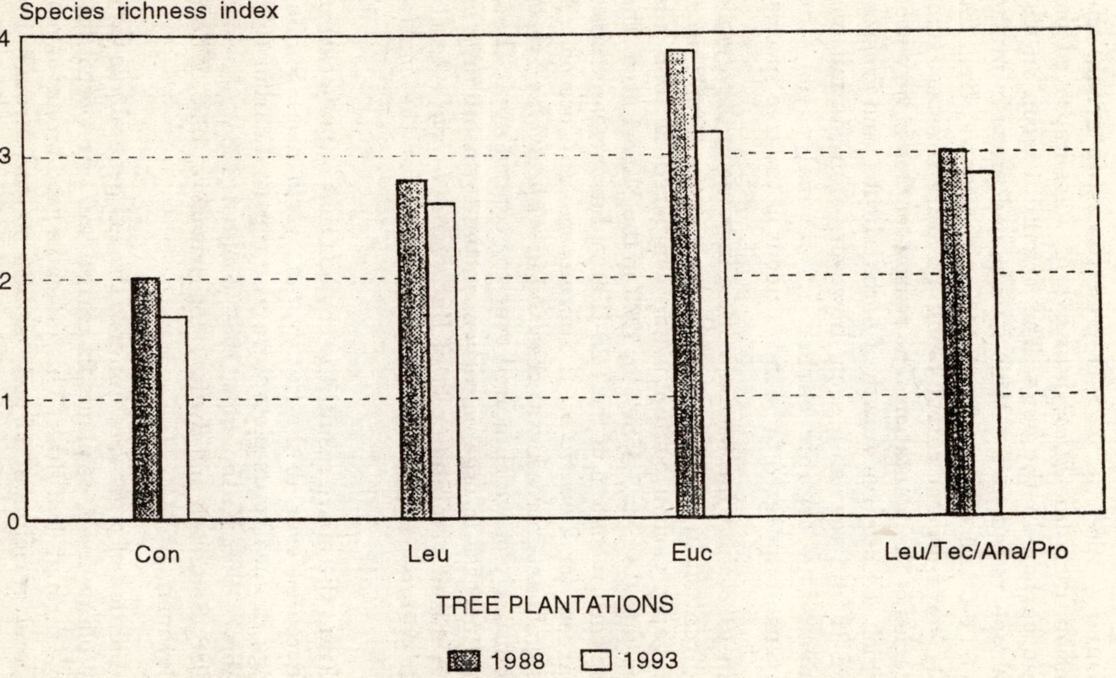

Figure 15. Species richness index values of typical plantations in mining area.

SOIL ORGANIC CARBON

The results are presented in Table 14.

During 1988, in TPS-II, and township areas, the plantation, including *Eucalyptus hybrid,* have improved the organic carbon in the soils. The organic carbon in the plantations raised in the Mine-I reclaimed areas is in the range 0.06-0.12% compared to 0.14% in the control area, the slightly lower organic carbon being a feature irrespective of the species of the plantation. The reason for this is very poor soil which discourages growth of herbs, shrubs, and grasses. In TPS-II and township area, however, afforestation has improved the organic carbon in several cases, the improvement appeared to be independent of the tree species.

In 1993, there has been a marked increase in the organic carbon content in all types of plantations in the Mine-I area. The organic carbon content when compared to 1988 has increased to 0.419 to 1.4589% in 1993 in the Mine-1 area. The organic carbon content of the TPS-II area, has increased since 1988 except for a *Acacia auriculiformis* monoculture plot. A very significant increase was observed in a *Eucalyptus hybrid* monoculture plot which scored over the control area. There was an increase in organic carbon in the township area as well. A mixed plantation plot of *Eucalyptus hybrid—Acacia auriculiformis—Prosopis juliflora*—scored over the control area.

From the above result it is obvious that organic carbon has increased even in the *Eucalyptus hybrid* plots. Similar findings in eucalyptus plots were also reported earlier (Gill and Abrol, 1966; Kushalappa, 1984; Singhal, 1984; Abbasi *et al.,* 1988; Sanginga and Swift, 1992; Srivastav, 1993; Wilson and Bowman, 1994).

Addition of organic cabron into the soil generally is done by the plants itself, as plants fix carbon and transport it into the soil through leaf litter. This indicates good breakdown of litter and enrichment of soil organic matter through the plantation. Another reason for improvement in organic matter could be the non-interference of humans into the plantation to collect forest products especially leaf litter. Since this will

Table 14. Soil organic carbon in the afforested and in the open ('Control') areas: Illustrative values*

Species	*Location*	*Organic carbon (%)*		*% increase*
		1993	*1988*	*in the organic carbon content*
TPS-II				
None	Near helipad	0.4626	0.09	+0.3726
E. hybrid	"	0.1963	0.15	+ 0.0463
A. auriculiformis	East of TPS-II	0.1262	0.16	-0.0338
E. hybrid	Near Checkpost	0.5327	0.15	+ 0.3827
Township				
None	Block 7	0.8692	0.16	+ 0.7092
E. hybrid	Block 16	0.3084	0.18	+ 0.1284
E. hybrid and A. auriculiformis	Block 6	0.4065	0.21	+ 0.1965
E. hybrid and A. auriculiformis	Block 5	0.4766	0.28	+ 0.1966
E. hybrid, A. auriculiformis and P. juliflora	Block 16	1.089	0.17	+ 0.919

* An improvement has occurred in all the Mine-I plots: the organic carbon % in 1988 was 0.06 to 0.12 whereas in 1993 it is 0.41 to 1.10%.

enhance the chance for more litter to decompose effectively. Thus, more humus is produced. It is also reported by Singhal *et al.* (1975) and Singhal (1984) that organic matter is more in eucalyptus plantation, as humification is more rapid under eucalypts as well as the litter of eucalypts is more hydrolysable than the litter of other trees. One more reason that could be attributed is that, enhancement of organic carbon might have occurred mainly due to the factor that utilization is less as compared to its production, which is also a general phenomenon observed in almost all eucalypts soil (Jha and Chimwal, 1993).

Hence, in the present study *Eucalyptus hybrid* has not in any way harmed the litter decomposition when compared to other plantation. As such, in the present study *Eucalyptus hybrid* has not depleted soil organic carbon.

SOIL NUTRIENTS (NPK)

The results are presented in Table 15.

In 1988, no significant difference was found between plots under *Eucalyptus hybrid* and plots under other single or mixed tree species. In 1993, the NPK status of control area in Mine-I is better than the plantation areas. Out of the six plots studied for NPK, the nitrogen content of the soil had increased in three plots since 1988. Fall of nitrogen was visible in one *Dalbergia sissoo* plot, one *Acacia auriculiformis* and one *Eucalyptus hybrid* monoculture plot. There was almost no change in the phosphorous content. However, there was a fluctuation of potassium in the plantation.

In TPS-II area, the phosphorous and potassium content was more in the control area than in plantation areas. There was an increase of nitrogen in *Eucalyptus hybrid* plantations since 1988. Potassium has increased significantly in one of the *Eucalyptus hybrid* plots. The other plots regardless of the type of plantation, showed a decline in phosphorous and potassium contents *vis-a-vis* 1988.

In the township area the NPK content do not show any rise.

In most cases phosphorous and potassium have gone down whereas nitrogen has increased. Potassium levels were lower in *Eucalyptus hybrid* site because potassium is known to be subject to high uptake by *Eucalyptus hybrid* species (Sanginga *et al.,* 1991; Tiwari and Mathur, 1983). Potassium is also absorbed by plants in relatively high amount.

Phosphorous and potassium are generally lower as they are taken up from the soil and get locked in the plant biomass with only a fraction coming back to the soil through leaf litter. Also, the amount of nutrient released from decomposing materials over a given time is a function of the quality and quantity of litter, and the soil physico-chemical and biological environment (Swift *et al.,*1979). Mineralisation of phosphorous also is greatly dependent on the C:N:P ratio of the organic matter.

Table 15. **NPK (Nitrogen, Phosphorus and Potassium) in the afforested and in the open (control) areas: Illustrative values**

Species	*Location*	*Nitrogen (kg/acre)*			*Phosphorus (kg/acre)*			*Potassium (kg/acre)*		
		1993	*1988*	*kg/acre increase in the nitrogen content*	*1993*	*1988*	*kg/acre increase in the phosphorus content*	*1993*	*1988*	*kg/acre increase in the potassium content*
1	**2**	**3**	**4**	**5**	**6**	**7**	**8**	**9**	**10**	**11**
Mine-I										
None	Near plot 4	91.0	unknown	91.0	0.22	unknown	0.22	122.20	unknown	122.20
E. hybrid	S_5 High dump	8.40	14.0	-5.6	<0.220	trace	No improve ment or same	17.50	18.0 18.0	-0.50
D. sissoo	"	21.70	14.0	+ 7.7	<0.220	"	"	22.5	18.0	+4.5
A. anriculiformis	"	24.5	16.8	+ 7.7	<0.220	"	"	23.5	21.0	+ 2.5
A. auriculiformis and E. hybrid	"	16.10	9.8	+6.3	<0.220	"	"	15.0	18.0	-3.0
D. sissoo	North dump	7.00	16.8	-9.8	<0.220	"	"	22.50	15.0	+7.5
A. auriculiformis	"	6.3	12.6	-6.3	0.220	"	Slight improvement	31.50	12.6	+18.9

Table 15. Contd.

1	2	3	4	5	6	7	8	9	10	11
TPS-II										
None	Near helipad	112.0	-	-	0.66	-	-	116.00	-	-
E. hybrid	"	112.0	44.8	+6.7.2	0.44	1.08	-0.64	100.50	120.0	-19.5
E. hybrid	Near checkpost	70.0	44.8	+25.2	0.22	1.15	0.93	100.00	15.0	+85.0
A. *auricu liformis*	East of TPS-I	15.4	56.0	-40.6	<0.22	Trace No change	18.3	33	–14.7	
Township None	Block 7	98.0	-	-	0.44	-	-	40.90	-	-
E. hybri	Block 16	52.0	56.0	-4.0	0.22	1.6	-1.38	31.50	64.8	-33.33
E. hybrid and A. auricu liformis	Block 6	140.0	70.0	+70.0	0.44	2.4	-1.96	18.50	61.8	–43.3
E. hybrid, A. auriculi-formis and P. juliflora	Block 16	56.0	56.0	0.0	0.33	1.6	-1.27	21.00	65.4	-44.4

PH - Township 7.5 - 8.0.

pH of the soil also controls the availability of phosphorous. It is reported that, neutral soil pH condition enhances the availability of phosphorous in the soil (Narain *et al.,* 1990). Hence, in the present study the reason for the less availability of phosphorous in the soil could be due to the soil pH condition which is found to be in the range of 7.5-8.5.

Decrease in nutrient amount may be attributed to the recycline pattern of eucalypts where uptake is more due to its fact growing nature but return through leaf fall is less (Nandi *et al.,* 1991). These observations are consistant with the findings of numerous workers on the nutrient status of forest soils which state that a thickly forested soil is poorer in phosphorous and potassium compared to a similar but deforsted soil in a similar agro-climatic regime (George, 1977; Pathak *et al.,* 1984; Balagopalan and Jose, 1986; Clements and Krishnamurthy, 1986; Biswas and Mukherjee, 1987; Kushalapa, 1987; Srivastav, 1993).

Nitrogen in the soil has increased since several leguminous undergrowth species have come into the plantations and their number also have increased form 1988 to 1993, thereby enriching soil nitrogen by fixing it from the air (through the nitrate pathway utilised by nodules). Thus, inspite of more vegetation in 1993 compared to 1988, soil nitrogen is higher in most plantations. This could be because during the decomposition of the litter, most of the nitrogen released during decomposition is taken up by micro-organisms for synthesis of their cell substances (Soni and Jamaluddin, 1990). Hence, as decomposition proceeds the soil is enriched with nitrogen through the micro-organisms.

Further, *Eucalyptus hybrid* has not performed differently, or less gainfully, than other species on the soil structure and fertility. Similar findings on other eucalyptus plantations were also reported by Singhal *et al.* (1975), Singhal (1984) and Kushalappa (1985).

SOIL MOISTURE

The results are presented in Table 16.

During 1988, in the TPS-II, and township areas, the plantation including *Eucalyptus hybrid,* have improved the moisture in their soils when compared with the control plot. In the control Mine-I reclaimed area, there has been significant improvement in the soil moisture. In all, the moisture content in *Eucalyptus hybrid* plantation is either more that of other plantations or more or less equal.

Table 16. Soil moisture in the afforestated and in the open ('Control') areas: Illustrative Values

Zone		*Moisture %*		
Species	*Location*	*1993*	*1988*	*% increase in the soil moisture content*
Mine-I				
None	Near plot 4	3.59	2.6	+0.99
E. hybrid	plot 3	10.07	9.9	+0.17
E. hybrid	plot 11	7.76	9.6	-1.84
A. auriculiformis	plot 4	6.49	9.1	-2.61
L. leucocephala	plot 6	10.62	9.0	+1.62
E. hybrid and A. auriculiformis	plot 2	9.07	5.75	+3.32
TPS-II				
None	Near helipad	0.54	2.5	+1.96
E. hybrid	"	4.76	6.5	-1.74
E. hybrid	East of TPS-II	0.79	5.0	-4.21
A. auriculiformis	"	0.47	3.6	-3.17
Township				
None	Block 7	0.26	1.73	-1.47
E. hybrid and A. auriculiformis	Block 6	0.99	2.2	-1.21
E. hybrid, A. auriculiformis and P. juliflora	Block 16	0.12	5.1	-4.98
E. hybrid	Block 16	0.38	4.1	-3.72

In 1993 there has been an improvement in the soil moisture content in the Mine-I reclaimed area except two monoculture plots of *Acacia auriculiformis* and *Eucalyptus hybrid*. The mixed plantation of *Eucalyptus hybrid—Acacia auriculiformis* plot showed best improvement in moisture content in the Mine-I area when compared to others. All types of plantations had better moisture retaining capacity than the control area. In 1988, all the plots had recorded an increase in moisture content in all the three zones since plantation. In TPS-II and township area, there has been a fall in the moisture content in all types of plantation and even in the control area.

In the Mine-I area the reason for the drop in soil moisture content than the control area can be the large surface area which has more transpiration rate. The fall in moisture content in TPS-II and township area is not too significant. Since the moisture content in *Eucalyptus hybrid* plantation is comparatively high in 1988 and even in 1993 the loss in soil moisture is notice in almost all plantations irrespective of type of plantations. The fall in moisture content in noticed even in the control plot (Township). The fall in the moisture content could be the month of sampling. The moisture content of Mine-I area was measured in February 1993 and other areas in April 1993. The entire 1988 study was done during May-June. The ambient temperature and humidity which can strongly influence the soil moisture were different during these different sets of observations. Similar findings were also reported by Wilson and Bowman (1994). According to them the general pattern of soil moisture content in three savanna communities corresponds to the seasonal rainfall patterns with maxima towards the end of the wet season followed by rapid decline in the dry season. Therefore, it is difficult to make meaningful comparison between the 1988 and 1993 findings.

One conclusion though can be drawn unambiguously—*Eucalyptus hybrid* has not caused any different impact on soil moisture than *Acacial auriculiformis—Leucaena leucocephala* monoculture or other mixed cultures. This staement is supportive of the other works done before (Heith and Karschon, 1963; Dabral and Subba Rao, 1969; Dabral, 1970; Thomas *et al.*, 1972; Abbasi *et al.,* 1988).

ORNITHOLOGICAL AND LEPIDOLOGICAL SURVEYS

We recorded 52 species of birds and 21 species of butterflies (Tables 17 and 18). In an earlier study conducted by Bombay Natural History Society (BNHS, 1991) covering the same areas as done by us, but in the month of December, 66 species of birds were recorded.

The difference in the number of species between the 1993 study and the BNHS report could be due to the fact that the BNHS study was conducted during winter season (December) immediately following the south west monsoon, when the situation is ideal for migratory birds to visit—when more number of species are expected. In fact it is surprising that even during summer—when the present study was conducted as many 52 species of birds could be recorded.

The reason for the number of birds present along the plantations could be the presence of water bodies which in general attract the water birds like *Grey heron, Pong heron, Large egret etc.* Another factor could be the temperature of the place and the presence of shrubs and herbs inside the plantation areas, the fruits of which attract the avian fauna. Same is the case with butterflies where shrubs like jatropha, calotropis *etc* attract the butterflies. This study shows that even in a reclaimed land, birds and butterflies have been visiting contrary to popular notions, there by showing that *Eucalyptus hybrid* has not discouraged birds or butterflies. Similar findings were reported by Steyn (1977), Woinarski (1979) and Kondas (1986). However, the density and diversity of birds was lower than in a natural forest. This could be best explained by the statement made by Budowski (1984), when he state: *As to ecological deserts, it all depends on with what a eucalyptus plantation is compared. If it is with a nearby natural mixed forest, there is no doubt that the latter is much richer in fauna but if such a plantation is compared with a nearby scarcely covered slope as for instance a burnt over savanna, then it is highly probable that there is more animal life including nesting birds in the eucalyptus stand.*

Table 17. Birds found in the afforestated areas of Mine-I

Sl. No.	*Common Name*	*Scientific Name*
1	**2**	**3**
1.	Darter	*Anhing rufa*
2.	Little egret	*Egretta garzetta*
3.	Cattle egret	*Bubulus ibis*
4.	Grey heron	*Ardea cinerea*
5.	Pond heron	*Ardeola grayii*
6.	Little cormorant	*Phalacrocorax niger*
7.	Blue rock pigeon	*Columba linia*
8.	Common sand piper	*Tringa hypoleucos*
9.	Coot	*Fulica atra*
10.	Indian moorhen	*Gallinula chloropus*
11.	Red Wattled lapwing	*Vanellus indicus*
12.	Little ringed plover	*Charadrius dubuis*
13.	Ashy swallow shrike	*Art anus fuscus*
14.	Ashy wren warbler	*Prinia socialis*
15.	Black drongo	*Dicrurus adsinilis*
16.	Blue tailed Bee-eater	*Merops philippinus*
17.	Brahminy kite	*Haliastur indicus*
18.	Brahminy myna	*Sturnus pagodarum*
19.	Common hawk cuckoo	*Cuculus varius*
20.	White breasted king fisher	*Halcyon snyrnensis*
21.	Indian myna	*Aerido theres tristis*
22.	Golden oriole	*Oriolus oriolus*
23.	Grey patridge	*Francolinus pondicerianus*
24.	House crow	*Corvus splendens*
25.	Indian robin	*Saxicoloides fulicate*
26.	Jungle crow	*Covus macrorhynchos*
27.	Koel	*Eudynamys scolopacea*
28.	Large pied wagtail	*Motacilla moderaspatensis*
29.	Pied king fisher	*Ceryle rudis*
30.	Common pariah kite	*Milvus migrans*

1	2	3
31.	Purple moorhen	*Porphyrio porphyrio*
32.	Purple rumped sunbird	*Nectarinia Zeylanica*
33.	Purple sun bird	*Nectarinia asiatica*
34.	Red vented bulbul	*Pycnonotus cafer*
35.	River tern	*Sterna aurautia*
36.	Rose ringed parakeet	*Psittacula krameri*
37.	Small green bee eater	*Merops orientalis*
38.	Indian tree pie	*Dendrocitta vagabunda*
39.	Spotted dove	*Streptopelia chimensis*
40.	Tailor bird	*Orthotomus sutorius*
41.	Tickell's flower pecker	*Dicaeum crythrorshynchos*
42.	Small green king fisher	*Alcedo atthis*
43.	White browed bulbul	*Pyenonotus lutcolus*
44.	Yellow wattled lapwing	*Vanellus malabricus*
45.	Grey Drongo	*Dicrurus leucophaeus*
46.	Sky lark	*Alanda arvensis*
47.	Common swallow	*Hirurndo rustica*
48.	Large egret	*Ardea alba*
49.	Brown shrike	*Lanius cristatus*
50.	Common Indian nightjar	*Caprimulgus affinis*
51.	Crimson throated barbet	*Megalaina rubricapilla malabrica*
52.	White headed babbler	*Turdoides affinis*

Table 18. Butterflies found in the afforestated areas of Mine-I

Sl. No.	*Common Name*	*Scientific Name*
1.	Striped tiger	*Danaus genutia*
2.	Common wanderer	*Pareronia veleria*
3.	Pale glass blue	*Zizeoria maha*
4.	Yellow pansy	*Funonia wierta*
5.	Common leopard	—
6.	Blue pansy	*Funonia orithya*
7.	Danaid eggfly	*Hypotimnas misippus*

1	2	3
8.	Tawny coster	*Acraca terpsicose*
9.	Plain tiger	*Danaus chrysippens*
10.	Blue tiger	*Tircumals limnace*
11.	Crimson rose	*Pachliopta nector*
12.	Common rose	*Pachliopta aristocohiae*
13.	Leopard	*Phalanta phalantha*
14.	Lemon pansy	*Funonia lemonias*
15.	Peacock pansy	*Funonia almana*
16.	Gram blue	*Euchrysaps enefu*
17.	Common Acacia blue	*Surendra quercetorum*
18.	Common sailor	*Neptishylas spp*
19.	Common nawab	*Polyma thmas*
20.	Golden augle	*Caproma ransonnathi*
21.	Common grass yellow	*Eurema Hecabe*

6

Summary and Conclusions

1. About 180 species of undergrowth were recorded in 1993 compared to 128 species of undergrowth in 1988 in the afforested areas of NLC. Of these, 69 new species were recorded only in 1993. The species included seasonals/ perennials/legumes/non-legumes and herbs/ shrubs/ saplings/climbers/sedges.

2. There is a 63.3% increase in the number of undergrowth species in 1993 compared to 1988.

3. Many of the undergrowth species are medicinally important. Several others have recognised uses as ornamental, edible, aromatic, fodder, manurial or wood-yielding species.

4. There was richer and denser undergrowth in the plantations in the township area compared to the plantations of similar age in the West dump area.

5. Qualitative as well as computer-aided quantitative studies gave clear indication that *Eucalyptus hybrid* monocultures have recorded impressive improvements in undergrowth; easily matching the performance of *Acacia auriculiformis* or *Leucaena leucocephala* monocultures and polycultures.

6. The moisture content of Mine-I area was measured in February 1993 and 'other areas' in April 1993. The earlier study in 1988 was done during May-June. On comparison with other areas, it may be inferred that *Eucalyptus*

hybrid has not caused any different impact on soil moisture than *Acacia auriculiformis*/*Leucaena leucocephala* monoculture or other mixed cultures.

7. There has been an increase in the soil organic carbon in all types of plantations over the years. *Eucalyptus hybrid* has not depleted soil orgainc carbon.

8. There has been an increase in the nitrogen content in most of the plots. This could be due to the increase in number of leguminous undergrowth species which enrich the soil nitrogen content. However, there has been a decline in the soil phosphorus and potassium content. In this respect, too, *Eucalyptus hybrid* has not performed differently or less gainfully than other species.

9. Fifty two species of birds and twenty one species of butterflies were recorded in the plantation areas thus disproving the allegation that eucalypts drives away wild life.

10. The reasons for the absence of negative impacts of eucalyptus in NLC could be: The correct choice of species (hybrid of *Eucalyptus tereticornis* and *Eucalyptus globulus* commonly known as *Eucalyptus hybrid*), and non-interference in terms of removal of leaf litter or sustained growth extractions. Such interference, which is common elsewhere, leads to stress on the soil and is the factor behind the main accusation against eucalyptus.

Bibliography

Abbai, S.A., Soni, R., Redy, A.S., Karthikeyan, N., Hema, S. and Uma, G. (1988). Environmental Management of Energy Projects—*Impact of Eucalyptus on Afforestation in mining Areas.* Compleion Report, Pondicherry (central) University, Pondicherry: 36 pages.

Ackerson, R.C. (1980). Stomatal response of cotton to water stress and absensic acid as affected by water stress history. *Plant Physiology,* Vol. 65, pp. 455-459.

Alexander, T.G., Balagopalam, M., Thomas, P.T. and Marry, M.V. (1981). *Properties of soil under Eucalyptus.* Res. Report, (8) Kerala Reserach Institute, Peechi: 12 pages.

Al. Monsawi, A., and Al. Naib, F.A.G. (1975). Allelopathic Effects of *Eucalyptus miccotheca. 'J' of the University of Kuwait (Science),* Vol. 2, p. 59.

Attwill, P.M. (1966). The chemical composition of rainwater in relation to cycling of nutrients in mature Eucalyptus forest. *Plant and soil,* 24 (3), pp. 390-406.

Bahuguna, S. (1984). *'My experience of Eucalyptus'.* Presentation at workshop on social and economic Impact of Eucalyptus. Planning commission (May 21), New Delhi, India.

Balagopalan, M. (1986). *Eucalyptus in India—Past, Present and Future.* Scientific paper, (71), KFRI, Peechi.

Balapopalan, M. and Jose, A.I. (1986). *Distribution of Organic Carbon and different forms of nitrogen in a natural forest adjacent Eucalyptus Plantation at Arippa.* Kerala, KFRI, Peechi.

Bargali, S.S. and Singh, S.P. (1991). Aspects of productivity and nutrient cycling in an 8-year old eucalyptus plantation in a moist plain area adjacent to central Himalaya, India *Can. J. For. Res.,* Vol. 21, pp. 1365-1372.

Bhasa (1986). ***Eucalyptus in India-Past, Present and Future,*** Scientific paper, (77), KFRI, Peechi.

Bhumla (1984). ***Summary Record of the Workshop on Social and Economic Aspects of Eucalyptus.*** MLP Unit, Planning Commission No. pc (PO 9/39/84, MLP), 8 pages.

Biswas, T.S. and Mukherjee, S.K. (1987). *Soil Sciences.* Tata Mc. Graw Hill Publishing Company Limited, New Delhi.

Bombay Natural Histroy Society (cited as BNHS) (1991). ***Checklist of birds at NLC (Mines-I Afforestation),*** unpublished report, NLC, Tamil Nadu, India. .

Brown, K.W., Jordon, W.R. and Thomas, J. (1976). Water stress induced alternation of stommatal response to decrease in leaf water potential. ***Physiol. Plant,*** Vol. 37, pp. 1-5.

Budowski, G. (1984). ***Farm and Community Forestry.*** Gerald Foley and Bernard Geoffrey. Earthscan Energy Information Programme. Nataraj Publishers. Dehra Dun: pp. 236.

Centre for Advanced Research and Development (cited as CARD) (1987). ***Internal meteorological, air and water quality report,*** NLC, Neyveli, India.

Centre for Science and Environment (cited as CSE a) (1985). ***The state of India's Environment.*** The second citizens' report, New Delhi: p. 64.

Centre for Science and Environment (cited as CSE b) (1985). ***The state of India's Environment.*** The second citizens' report, New Delhi: p. 65.

Centre for Science and Environment (cited as CSE c) (1985). ***The state of India's Environment.*** The second citizens' report, New Delhi: p. 66.

Centre for Science and Environment (cited as CSE d) (1985). ***The state of India's Environment.*** The second citizens' report, New Delhi: p. 67.

Centre for Science and Environment (cited as CSE e) (1985). ***The state of India's Environment.*** The second citizens' report, New Delhi: p. 68.

Centre for Science and Environment (cited as CSE f) (1985). ***The state of India's Environment.*** The second citizens' report, New Delhi: p. 69.

Centre for Science and Environment (cited as CSE g) (1985). ***The state of India's Environment.*** The second citizens' report, New Delhi: pp. 67-68.

Chaturvedi, A.N. (1976). Eucalyptus in India. ***Indian Forester,*** 102 (1), pp. 57-63.

Chaturvedi, A.N. (1983). Eucalyptus for farming. ***Forest Bullettin,*** Vol. 48, U.P. Forest Department. Lucknow, India: p. 48.

Chinnamani, S., Gupta, S.C., Rege, N.D. and D. Thomas, P.K. (1965). Run-off studies under different forest cover in the Nilgiris. *Indian Forester,* 91 (8), p. 676-679.

Choudhury (1986). Eucalyptus in India-Past, Present and future. Scientific Paper, (71), KFRI, Peechi, Kerala.

Clements, V. and Krishnamurthy, R. (1986). ***Response of Eucalyptus 'hybrid' to Major Nutrient Elements.*** KFRI, Peechi, Kerala.

Cox, G.W. (1990). *Strip transect censusing of birds.* Laboratory manual of general ecology (Sixth ed), San Diego State University, U.S. A: pp. 64-68.

Dabral, B.G. and Subba Rao, B.K. (1969). Interception Studies in Sal (*Shorea robusta)* and Khair) (*Acacia catechu*) Plantations of New Forest. *Indian Forester,* Vol. 95, pp. 467-475.

Dabral, B.G. (1970). Preliminary Observations in Potential water Requirements in *Pinus roxburghii, Eucalyptus citriodora, Populus casale and Dalbergia latifolia. Indian Forester,* Vol. 96, pp. 775-778.

Davidson, J. (1985). *Setting aside the idea Eucalyptus are always bad.* Working paper, (10), Bangladesh.

de la Lama, G.G.C. (1984). Atlas del eucalipto. Ministerio de Agricultura, Institue Nacional de Investigaciones Agrorias e Instituto Nacional para la conservacion. *de la Naturaliya,* 5 (appendix), Madvid, p. 82.

Del Moral, R. and Muller, C.H. (1969). Fog Drip: A Mechanism of Toxin Transport from Eucalyptus globulus. *Bulletin of the Torreo Botanical Club,* Vol. 96, pp. 467-475.

Del Moral, R. and Muller, C.H. (1970). The allelopathic effects of *Eucalyptus camaldulensis American Midland Naturalist,* Vol. 83, pp. 254-282.

Dietz, J.M., Couto, E.A., Alfenas, C.A., Faccini, A. and da Silva, G.F. (1975). Efeitos de duas plantacoes de florestas homogeneas sobre populacoes de mamiferos pequenos. *Brasil Florestal,* Vol. 23, pp. 25-57.

Dinesh Kumar (1984). *Place of Eucalyptus in Indian Agroforestry Systems.* KFRI, Peechi, Kerala.

Evans, J. (1982). *Plantation Forestry in the Tropics.* The English Language Book Society and Clarendor Press, ELBS (Ed.), Oxford: 472 pages.

Foley, G. and Bernard, G. (1984). *Farm and Community Forestry.* Earthscan Energy Information Programme. Natraj publishers, Dehra Dun: 236 pages.

Friend, G.R. (1982). Mammal populations in exotic pine plantations and indigenous eucalypt forests in Gippsland, Victoria, *Australian Forestry,* 45(1), pp. 3-18.

Gamble, J.S. (1935). *Flora of the Presidency of Madras.* Vol. I, II and III. Published by Bishen, S. and Mahendra, P.S. Dehra Dun, India: 1989 pages.

George, M. (1977), *Organic Productivity and Mineral Cycling in Eucalyptus hybrid Plantations,* Ph.D. Thesis, Meerut University, Meerut.

George, M. (1979). Nutrient return by stemflow, throughfall and rainwater in a *Eucalyptus hybrid* plantation. *Indian Forester,* 105(7), pp. 493-499.

George, M. and Varghese, G. (1991). Nutrient cycling in Eucalyptus globulus plantations. III-Nutrients Retained, Returned, Uptake, and Nutrient cycling. *Indian Forester,* 117(2), pp. 110-116.

Gill, H.S. and Abrol, I.P. (1966). *Salt affected soils and their amelioration through afforestation.* Soil amelioration by tree. Commonwealth Sci. council. pp. 43-53.

Gupta, R.K. (1986). *Eucalyptus in India—Past, Present and Future.* KFRI, Peechi. Kerala.

Gurumurti, K. and Rawat, P.S. (1989). Time trend studies on biomass production in high density plantation of *casuarina equisetifolia.* Proc. of the National Seminar on Casuarinas (December 18 and 19). Tamil Nadu Forest Plantation Corporation and Institute of Forest Genetics and Tree breeding, Coimatore, Tamil Nadu, India: Session III. 3.

Hassankutty, A. (1986). *Eucalyptus*—Myths and Truths. KFRI, Peechi, Kerala.

Heith, D. and Karschon, R. (1963). *Interception of Rainfall by Eucalyptus in Israel.* Mat. Univ. Inst. Agri Ilanot and Land Development Authority, Kiriat, Hayim, pp. 7-24.

Heith, D. and Karschon, R. (1967). *The Water Balance of Plantation of Eucalyptus camaldulensis. Dehn. Contributions of Eucalyptus in Israel III,* Ilanot and Kiriat Hayim, Israel, pp. 7-34.

Jamet, R. (1975). Evolution des principles carocteristiques des sols des reboisements de Londima (congo), cashier ORSTOM, *serie Pedologie,* 8(3/4), pp. 235-253.

Jensen, A.M. (1983). *Shelter belt effects in tropical and temperate zones.* International Development Research Centre Manuscript Reports, IDRC-MR80e, 61 pages.

Jha, K.K. and Chimwal, C.B. (1993). Performance of different provinances of *Eucalyptus camaldulensis in* Tevai and its effect on soil properties. Indian Journal of Forestry, 16(2), pp. 97-102.

Jha, M.N. and Pande, P. (1984). Impact on growing Eucalyptus and soil monocultures on soil in natural soil area of Doon Valley. *Indian Forester,* 110(1), pp. 16-22.

Jha, P.S. (1984). *Farm Forestry Under Attack.* The Times of India (New Delhi), August 2, 3, and 8.

Jocque, C.A. (1981). *A terrestrial baselines study of the Viphya Pulpmill Project Area.* FAO, internal report.

Joshie, P. and Narain, P. (1994). Vegetation characteristics and nutrient composition of underwood flora in Sal, Eucalyptus and Brushwood forest watersheds of Doon Valley. *The Indian Forester,* 120(4), pp. 331-332.

Joyce, C. (1988). The tree that caused a riot. *New Scientist,* 117(1600), pp. 54-56.

Kaikini, N.S. (1967). *Proceedings of the 11th Silvicultural Conference.* (1), Forest Research Institute and Colleges, Dehra Dun: 93 pages.

Karanth, U. and Singh, M. (1983). *Dry Zone Afforestation and Its Impact on Blackbuck Population.* Proceedings of the Centenary Seminar of the BNHS, Bombay.

Kawahara, T., Kanazawa, Y. and Sakurai, S. (1981). Biomass and net production of man-made forest in the Philippines. *Journal of the Japanese Forestry Society,* 63(9), pp. 320-327.

Kedarnath, S. (1986). *Eucalyptus in India—Past, Present and Future.* India, ed., KFRI, Peechi.

Khan, M.S. (1959). *Dry zone afforestation in Andhra Pradesh.* Technical paper, Dry zone afforestation symposium (January-February), (3), Forest Research Institute, Dehra Dun: pp. 1-11.

Kondas (1986). *Eucalyptus in India—Past, Present and Future.* KFRI, Peechi, Kerala.

Kushalappa, K.A. (1984), *Nutrient status in "Eucalyptus hybrid" monoculture.* Abstract of Paper for National Seminar on *Eucalyptus,* January 30-31, Kerala, India.

Kushalappa, K.A. (1985). Nutrient status in *Eucalyptus hybrid* monoculture, *Indian Journal of Forestry,* 8(4), pp. 269-673.

Kushalappa, K.A. (1987). Productivity of *Eucalyptus hybrid* Under Different Ecosystems in Karnataka. *My Forest,* 24(4), pp. 217-226.

Lamb, D. (1973). *A suspected phosphorous and potassium deficiency in the tropical eucalypt—E. deglupta.* Paper for FAO/IUFRO International symposium of Forest Fertilization, December 3-7, Paris.

Lewis, J.K. (1970). *Primary production in grass land eco-system.* U.S./ I.B.P. Ovanland Ecosystem. Supplement. Ed. R.L. Dix and R.C. Beidleman.

Lima, W.P. and O' Loughlin, E.M. (1984). *The hydrology of Eucalypt forests in Australia*—a review. Submitted for publication in IPEF, Piracicaba, Brazil.

Mahashweta Devi (1983). 'Why Eucalyptus?' *Economic and Political Weekly.* 78(32), pp. 1379-1381.

Mathur, H.N. and Soni, P. (1983). Comparative account on undergrowth under Eucalyptus and Sal in three different localities of Doon Valley, *Indian Forester,* 109(12), pp. 882-890.

Mathur, H.N., Naveen, J. and Sajwan, S.S. (1980). Ground cover and undergrowth in Eucalyptus, Brushwood and Sal forest—an ecological assessment. *Van vigyan,* 18 (3 and 4).

Mathur, H.N., Ram, B. and Joshie, P. (1976). Effect of clear felling and reforestation on run-off and peak rates in small watershelds. *Indian Forester,* 102(4), pp. 219-225.

Mc Naughton, S.J. (1967). Relationship among functional properties of california grassland. *Nature,* Vol. 216, pp. 168-169.

Nandi, A., Basu, P.K. and Banerjee, S.K. (1991). Modification of some soil properties *Eucalyptus* species. *Indian Forester,* 117(1), pp. 53-57.

Narain, P., Singh, R. and Singh, K. (1990). Influence of Forest covers on Physico-chemical and site characteristics in Doon Valley. *Indian Forester,* 116(11), pp. 901-916.

Neginhal, S.G. (1980). Ecological impact of afforestation at the Ranibennur Black buck Sanctuary. *Journal of the Bombay Natural History Society,* Vol. 75 (suppl), pp. 1254-1258.

Newmann, F.G. (1979). Beetle communities in eucalypts and pine forests in north-eastern victoria *Australian Forest Research,* 9(4), pp. 277-293.

Neyveli Lignite Corproation (cited as NLC) (1988 a). *In Pursuit of Excellence—An Afforestation Experience.* NLC, Neyveli: 58 pages.

Neyveli Lignite Corporation (cited as NLC) (1988 b). *Environmental Management Plan For Mines I (Expansion), Neyveli.* Min. Mec Consultancy Private Limited, New Delhi. Vol. I and II.

Pal, M. and Ratwri, D.P. (1991). Growth, Biomass production and Dry matter distribution pattern of *Eucalyptus hybrid* growth in an Energy plantation. *Indian Forester,* 117(3), pp. 187-192.

Pande, M.C., Tandon, V.N. and Negi, M. (1986). Biomass production and its distribution in an age series plantations of *Eucalyptus hybrid* and *Acacia auriculiformis* in Bihar. *Indian Forester,* 112(11), pp. 975-985.

Pande, M.C., Tandon, V.N. and Rawat, H.S. (1987). Organic matter production and distribution of nutrients in *Eucalyptus hybrid* plantation ecosystems in Karnataka. *Indian Forester,* 113(11), pp. 713-724.

Pathak, A.N., Harishankar and Mukherjee, P.K. (1984). A Study of the Physical and Chemical Properties of the Soil Under Cultivation of Forest Covers. *Indian Forester,* 90(3), pp. 171-175.

Penfold, A.R. and Willis, J.L. (1961). *The Eucalyptus—botany, cultivation, chemistry and utilisation.* Teonard Hill, London.

Poore, M.E.D. and Fries, C. (1987). *The Ecological Effects of Eucalyptus.* Natraj Publishers, Dehra Dun: 98 pages.

Prasad, K.G., Singh, S.B., Gupta, G.N. and George, M. (1985). Studies on changes in soil properties under different vegetation. *Indian Forester,* 111(10), pp. 794-801.

Pryor, L.D. (1976). *The biology of Eucalyptus.* Edward Arnold, London: 77 pages.

Quereshi, I.M. (1967). *The concept of Fast Growth in Forestry and Place of Indigenous Fast Growing Broad Leaved Species.* Proceedings of the 11th Silvicultural Conference, (2), Forest Research Institute and Colleges, Dehra Dun: 505 pages.

Raison, R.J. and Crane, W.J.B. (1981). *Nutritional costs of shortened rotations in plantation forestry.* 17th IUFRO World Congress. Kyoto-proceedings Div. I. pp. 63-72.

Rajvanshi, A., Soni, S., Kuketi, U.D. and Srivastava, M.M. (1983). A comparative study of undergrowth of sal forest and Eucalyptus plantation at Golatapper—Dehra Dun during rainy season. *Indian Journal of Forestry,* 6(2), pp. 117-119.

Rao, R. (1995). *The Eucalyptus controversy.* A special article to mark the world forestry day. March 21, vol. 09. Centre for

Environment Education—News and Features Service (CEE-NFS), Ahmedabad, India: 4 pages.

Rice, E.L. (1979), Allelopathy—an update. *Bot. Rev.*, 45, pp. 15-109.

Samraj, P., Chinnamani, S. and Haldorai, B. (1977). Natural verus man-made forest in Nilgiris with special reference to run-off, soil loss and productivity. *Indian Forester,* 103(7), pp. 460-465.

Sanginga, N. and Swift, M.J. (1992). Nutritional effects of Eucalyptus litter on the growth of maize (zeamays). *Agriculture, Ecosystems and Environment,* Vol. 41, pp. 55-65.

Sanginga, N., Gwaze, D. and Swift, M.J. (1991). Nutrient requirements of exotic tree species in Zimbabwe, *Plant soil,* Vol. 132, pp. 197-205.

Saxena, N.C. (1992). Farm Forestry and Land-use in India: Some Policy Issues. *Ambio,* (21)(6), pp. 420-425.

Shankaran (1995). *Productivity study of eucalyptus at Neyveli Lignite Corproation, Tamil Nadu, India* (Unpublished internal report).

Shanmungan. S. (1986). *Eucalypts in India—Past, Present and Future.* KFRI, Peechi, Kerala.

Shiva, V. and Bandyopadhyay, J. (1987). *Ecological audit of Eucalyptus Cultivation.* Research Foundation for Science and Ecology, Dehra Dun: 74 pages.

Shyamsunder, S. (1983a). *The Last Word on Eucalyptus.* Indian Express (December 7), Bangalore, India.

Shyamsunder, S. (1983 b). *Some Aspects of Eucalyptus hybrid.* Karnataka Forest Department, Bangalore; also letter from Shyamsunder, S. to Prof. M.D. Najundaswamy, President, Karnataka State Farmers Association, dated August 20.

Shyamsunder, S. (1986). *Eucalypts in India—Past, Present and Future.* KFRI, Peechi, Kerala.

Singhal, R.M. (1984). *Effect of growing Eucalyptus on the status of soil organic matter on Dehra Dun forests.* Abstract of Paper for National Seminar on Eucalyptus, January 30-31, Kerala, India.

Singhal, R.M., Banerjee, S.P. and Pathak, T.C. (1975). Effects of Eucalyptus monoculture on the status of soil organic matter in natural Sal (*Shorea robusta*) Zone in Doon Valley. *Indian Forester,* 101 (12), pp. 730-737.

Singh, R.P. (1984). Nutrient cycle in *Eucalyptus tereticornis* Smith Plantations. *Indian Forester,* 110(1), pp. 76-85.

Singh, S.B., Pramode, K. and Prasad, K.G. (1993). Potential water requirements of Eucalyptus—A preliminary study. *Indian Forester,* 119(7), pp. 549-553.

Soil analysis procedure (1988). Soil testing laboratory, Sugarcane research station, Cuddalore, Tamil Nadu, India: pp. 3-21.

Somachai, T., Pongsak, S. and Kyoji, Y. (1991). Litterfall and productivity of *Eucalyptpus camaldulensis* in Thailand, *Journal of Trophical Ecology,* Vol. 7, pp. 275-279.

Soni, K.K. and Jamaluddin (1990). Eucalyptus litter decomposition in tropical dry deciduous forest of Madhya Pradesh. *Indian Forester,* 116(4), pp. 286-291.

Srivastav, A.K. (1993). Change in physical and chemical properties of soil in irrigated Eucalyptus plantation in Gujarat State. *Indian Forester,* 119(3), pp. 226-231.

Stein, A.H. (1952). Nota Sobre los resultados obtenidos en otros paises en las experiencias acerca de la influencia del *Eucalptus* Sobre la Cubierta forestal de las hoyas hidrograficas Y sobre el mejoramiento del Suelo con su cylicacion a la misma materia en chile. Paper from "Mision Forestal de la FAO", No. 9. Santiago de chile.

Steyn, D.J. (1977). *Occupation and Use of The Eucalyptus Plantation in the Tzaneen Area by Indigenous Birds.* South Africa Forestry Tour No. 1005-60.

Story, R. (1967). Pasture patterns and associated soil water in partially cleared woodland. *Australian Journal of Botany,* Vol. 15, pp. 175-187.

Swami Rao, N. and Chandrasekhara Reddy, P. (1984). Studies on the inhibitory effects of *Eucalyptus hybrid* leaf extracts in the germination of certain food crops. *Indian Forester,* 110(2), pp. 208.

Swift, M.J., Heal, D.W. and Anderson, J.M. (1979). *Decomposition in Terrestrial Ecosystems.* Black well Scientific Publications, Oxford.

Thomas P.K., Chandrasekhar, K. and Haldorai, B. (1972). An estiamte of transpiration by *Eucalyptus globulus* from Nilgiris watershed. *Indian Forester,* 98(2); pp. 168-172.

Tiwari, K.M. and Mathur, R.S. (1983). Water consumption and nutrient uptake by Eucalyptus. *Indian Forester,* 109(12), pp. 851-860.

Tiwari, K.M. (1983). Soil and water conservation—need for better management in the country. *Indian Forester,* 109(11), pp. 775-780.

Williams, G. (1987). *Techniques and Field Work in Ecology.* Bell and Hyman Limited Publications, Denmark House, London: 156 pages.

Williams, J.E. (1990). The importance of herbivory in the population dynamics of three sub-alpine eucalypts in the Brin da bella Range, South-east Australia. *Aust. J. Ecol.,* Vol. 15, pp. 51-55.

Wilson, B.A., and Bowman, D.M.J.S. (1994). Factors influencing tree growth in tropical Savanna: Studies of an abrupt Eucalyptus boundary at Yopilika, Mel ville Island, Northern Australia. *Journal of tropical Ecology,* Vol. 10, pp. 103-120.

Wilson, S.D. and Zammit, C.A. (1992). Tree litter and the lower limits of subalpine herbs and grasses in the Brindobella Range, ACT. *Australian Journal of Ecology,* Vol. 17, p. 321-327.

Woinarski, J.K.Z. (1979). Birds of Eucalyptus Plantation and Adjacent Natural Forest. *Australian Forestry,* 42(4), pp. 243-247.

Yegnaswami, A. (1961). *Trials with Eucalyptus in Madras state.* Proc. 10th All Indian Silricultural Conference, Dehra Dun: pp. 601-609.

Index

F

G

H

I

K

L

M

N